D1605433

Ecological Studies

Analysis and Synthesis

Volume 33

Wolfgang Böhm

Methods of Studying Root Systems

With 69 Figures

Springer-Verlag Berlin Heidelberg New York 1979

Dr. WOLFGANG BÖHM
Institut für Pflanzenbau und Pflanzenzüchtung
der Universität Göttingen
v.-Siebold-Straße 8, 3400 Göttingen/FRG

ISBN 3-540-09329-X Springer-Verlag Berlin Heidelberg New York
ISBN 0-387-09329-X Springer-Verlag New York Heidelberg Berlin

Library of Congress Cataloging in Publication Data. Böhm, Wolfgang, 1936–. Methods of studying root systems. (Ecological studies; v. 33). Bibliography: p. Includes indexes. 1. Roots (Botany)-Research. 2. Botany-Ecology-Methodology. I. Title. II. Series. QK644.B63. 581.1. 79-9706

Typesetting, printing, and binding: Brühlsche Universitätsdruckerei, Lahn-Gießen.
2131/3130—543210

This book is dedicated to
John Ernest Weaver
(1884–1966)
pioneer in root ecology

Preface

Root research under natural field conditions is still a step-child of science. The reason for this is primarily methodological. The known methods are tedious, time-consuming, and the accuracy of their results is often not very great. Many research workers have been discouraged by doing such root studies.

The need for more information on the development and distribution of plant roots in different soils under various ecological conditions is, however, obvious in many ecological disciplines. Especially the applied botanical sciences such as agriculture, horticulture, and forestry are interested in obtaining more data on plant roots in the soil.

This book will give a survey of existing methods in ecological root research. Primarily field methods are presented; techniques for pot experiments are described only so far as they are important for solving ecological problems. Laboratory methods for studying root physiology are not covered in this book.

Scientific publications on roots are scattered in many different journals published all over the world. By working through the international root literature I found that about ten thousand papers on root ecology have been published at the present. This is not very much compared with the immense literature on the aboveground parts of the plants, but is, however, too much to cite in this book. Therefore, mainly papers that contribute to methodology have been cited. As traditional methods are still in use at the present time, older literature has also been referenced. This should be not least a debt of honour to the research workers in the pioneer years of root ecology. We should not forget their contributions, often obtained with the simplest equipment, to methodological progress.

The author is not familiar with the Russian language. The user of this book must excuse the fact that only few original Russian publications are mentioned in the literature index. Fortunately many publications of Russian scientists have been translated into English or German, and a complete survey of the root-study methods used in the Soviet Union is accessible in English through the book of Kolesnikov (1971).

I know that to write a book on methods is an invitation to criticism. It must, however, be taken into consideration that such a book should be neither too voluminous nor too expensive. Compromises must be made between the wishes of the author and those of the publisher, and therefore all the methods cannot be described in detail.

While writing this book I had the feeling that experts from neighbouring ecological sciences were looking over my shoulder and some seemed to be more qualified than I to write one chapter, or the other. These experts, especially, but also everybody who uses this book, are invited to make improvements. In spite of omissions and shortcomings, I hope with this book to give research workers in many areas of botany, agriculture, horticulture, forestry and other ecological sciences a helpful guide to the various methods of ecological root research.

I am most grateful to many colleagues all over the world who helped me with many suggestions and criticism in preparing the manuscript. I wish to record my thanks to D. Atkinson (England), W.A.P. Bakermans (Netherlands), G.H. Chilvers (Australia), M.C. Drew (England), W.M. Edwards (USA), W. Ehlers (Germany), F.B. Ellis (England), J.B. Fehrenbacher (USA), V. Haahr (Denmark), J.A. Heringa (Netherlands), A.R.G. Lang (Australia), W.H. Lyford (USA), H. Mohr (Germany), A. Reijmerink (Netherlands), H.C. De Roo (USA), R. Steinhardt (Israel), D. Tennant (Australia), and P.J. Welbank (England).

I especially want to thank G.M. Aubertin (USA) for the use of his excellent root bibliography and for his valuable criticism and rewriting in parts of the text. In a similar way I thank H.M. Taylor (USA) for his advice and critical improvements by reviewing main parts of the text and for his stimulation to write this book about methods. I very heartily thank K. Baeumer (Germany) for his encouragement during preparation of the text of the book and U. Köpke (Germany) who contributed valuable new methodological findings and gave critical advice. Thanks also to the numerous people at the university libraries at Göttingen (Germany) and Ames, Iowa (USA) for proving much of the literature.

I thank all the colleagues and editors of journals for permission to use photos and figures. I am indebted to Mrs. Jean von dem Bussche for revising the entire English text, Miss E. Müller for her typewriting, Miss S. Eigenbrodt for reading the proofs, and Miss E. Schuhmacher for her help in details of publication. I heartily thank all other friends and collaborators for their help. Thanks also to the Deutsche Forschungsgemeinschaft, who granted me financial support to study ecological root research in the USA.

Many thanks also to O.L. Lange (Germany) for his worthwhile help and for publishing the book in the series *Ecological Studies*, and to the publisher K.F. Springer, for the excellent cooperation during the preparation of the book.

Göttingen, May 1979 WOLFGANG BÖHM

Contents

1. Root Ecology and Root Physiology

The rapid development of root research in the last few years is leading more and more to a specialization in this area of science. Root ecology and root physiology are now the two main fields in root research, as was made strikingly clear at the Second International Root Symposium 1971 in Potsdam (German Democratic Republic) under the title *Ecology and Physiology of Root Growth* (Hoffmann, 1974a).

The aim of root ecology is to investigate the influence of environmental factors on the development of plant root systems. To attain this aim, the first task for a root ecologist will be to determine in which soil layer at which time how many roots of the plants are growing, and to obtain data on the dynamic development of the root systems. This fundamental work of ecological root research can be described by the terms root geography or rhizography (Lyford, 1975).

Obtaining root data in the field can, however, only provide the basic information for a root ecologist. To interpret his data, he needs information on the environmental factors which influence root growth; therefore root-study methods must always be combined with other ecological investigations. The period where root data were collected without at least simultaneous investigations of the soil profile in which the roots were growing should be over.

Important ecological factors which influence root growth are bulk density, strength, water, air and nutrients in the soil. The dependence of root growth upon these and other ecological factors is summarized by Weaver (1926), Rogers (1939a), Goedewaagen (1942), Troughton (1957, 1962), Kutschera (1960), Weller (1965), Pearson (1966), Danielson (1967), Köstler et al. (1968), Sutton (1969), Kolesnikov (1971), Trouse et al. (1971), and Cooper (1973). In these publications the field of root ecology covered by research workers from various ecological disciplines is outlined in detail.

The other pillar of root research is root physiology. This field is still an integral part of plant physiology, and this must also be the correct area for this kind of research in the future. The root physiologist is looking for physiological processes in the roots. His field of research is, for example, problems of cell division in root tips or transport mechanisms of ions in young tree roots.

A clear separation between root ecology and root physiology cannot and should not be made. In every case root growth is governed by external and internal factors. Therefore the root ecologist needs results from the root physiologist for interpreting his root data. On the other hand, the root physiologist must look at the results of ecological root research to see if his laboratory results are valid under field conditions.

2. General Survey of Root-Study Methods

2.1 Historical Development

Systematic studies on root systems were started in the eighteenth century. With a simple excavation technique Hales (1727) dug out root systems of cultivated crops and determined their morphology, as well as their weight and length. Du Hamel Du Monceau (1764/65) excavated tree root systems. Then for over one hundred years no important root research was documented.

Due to the increasing use of mineral fertilizer on agricultural land in the second part of the nineteenth century, more and more agronomic scientists became interested in root studies, so that most root excavations were made on agricultural plants. Pioneer research in this field was done in Germany (Thiel, 1870; Fraas, 1872; Werner and Schultz-Lupitz, 1891; Kraus, 1892–96, 1914). Schubart (1855, 1857) washed out root systems of cereal plants at a profile wall with water pressure from a garden hose.

At the same time many scientists used containers for studying root growth (Nobbe, 1862; Hellriegel, 1883; von Pistohlkors, 1898; Schulze, 1906, 1911/14). Sachs (1873) introduced the technique of observing roots behind glass panels as an in situ method, and at the beginning of the twentieth century two small root laboratories were constructed in Germany (Kroemer, 1905).

Considerable stimuli for further development of root study methods came from American scientists. King (1892) enclosed a large soil monolith within a metal cage, pushed wire rods through the cage, and after washing the soil away with water, observed the plant roots maintained nearly in their natural position. This method was improved by Goff (1897) to the classical needleboard method, later used on a large scale by Rotmistrov (1909), the leading root scientist in Russia at that time.

Cannon (1911) made his intensive root studies on desert plants using only the excavation technique. It was, however, the American plant ecologist J.E. Weaver (1884–1966) who developed the simple excavation technique with garden tools into a recognized scientific method. In his most important book, *Root Development of Field Crops* (1926), he described his techniques in detail. Weaver's classical excavation method was the prevalent method for studying root systems all over the world until the middle of this century. Weaver, professor at the University of Nebraska, did most of his root studies in the American Midwest prairie. Beside excavation, he used several other methods, and with his root research he influenced scientists all over the world; but his contribution to the development of modern root ecology was more appreciated abroad (Kmoch, 1957) than in his own country.

Weaver used the excavation method predominantly for root studies on prairie plants. Later, excellent work was done with this method by Laitakari (1929a, 1935), and Krauss et al. (1934) with forest trees, by Rogers and Vyvyan (1928, 1934) with orchard trees and by numerous other research workers with various other plant species.

Oskamp and Batjer (1932) demonstrated that by exposing and counting the number of roots at a smoothed profile wall, their distribution in a soil profile can be determined very easily. This profile wall method was at first used only for studies of tree roots, but thirty years later was also successfully applied to annual plants. Pavlychenko (1937a), a pupil of J.E. Weaver, published his famous technique for washing out complete root systems from giant monolith blocks. Conrad and Veihmeyer (1929) proposed the determination of the water content in soil samples as an indirect method for root studies.

Beside these methods, the sampling of soil-root samples by augers or other tools with an associated washing out of the roots from the soil and determining root weight was used more and more by scientists. In Germany, where ecologial root research was closely connected with humus research, most of the intensive studies have been done with this sampling technique (Günther, 1951; Könnecke, 1951; Könekamp, 1953; Köhnlein and Vetter, 1953).

The introduction of radioactive tracers into ecological root research by Hall et al. (1953) and other scientists seemed to be an immense step forward. However, enthusiasm has now declined, as the tracer methods can mainly be used only to supplement direct study methods. Working with isotopes has been more successful in studies of problems of root physiology than with root ecology.

Until the end of the second decade of our century knowledge of the root systems itself was the main research aim in field studies. Since then, more and more, the ecological conditions where the roots are growing have also been investigated simultaneously. This has led to real ecological studies and the basis of modern root ecology was given.

The need for more data on dynamic root development led in the last decade of our century to the construction of modern root laboratories, underground walkways, in which the roots can be observed behind glass panels. Since the construction of the prototype 1962 in East Malling, England (Rogers, 1969), about fifteen such facilities, also termed rhizotrons, have been built all over the world.

Today most of these root study methods are still in use by scientists. Weaver's classical excavation method was also brought to new bloom by Kutschera in Austria by her comprehensive root atlas (Kutschera, 1960). Other methods have been mechanized, for example the techniques of taking soil cores.

Because descriptions of methods and improved techniques for ecological root studies are scattered in various journals around the world, in recent years some short methodological reviews have appeared (Troughton, 1957; Shalyt, 1960; Krasilnikov, 1960; McKell, 1962; Carlson, 1965; Weller, 1965; Röhrig, 1966; Köstler et al., 1968; Lieth, 1968; Newbould, 1968; Yorke, 1968; Hoffmann, 1974b; Böhm, 1978a). Schuurman and Goedewaagen, the leading root research workers from the Netherlands, published in 1965 (2nd ed. 1971) a short manual of the methods which they used in their studies. Kolesnikov, the leading root ecologist in the Soviet Union, gave a survey of the main study methods used in his country for orchard trees (1971) and for forest trees (1972). The proceedings of the First International Root Symposium in Leningrad (USSR Academy of Sciences, 1968) also gave a survey of some common root-study methods.

2.2 Principles of Classification

A unique classification of existing root-study methods on a systematic basis is not possible because several methods, different in principle, have certain similar features. Therefore the grouping of the methods is more or less a personal decision.

In the simplest way a classification into old and new, or direct and indirect, or field and container methods can be made. In this book the methods are grouped similarly to the classification of Kolesnikov (1971):

1) Excavation methods
2) Monolith methods
3) Auger methods
4) Profile wall methods
5) Glass wall methods
6) Indirect methods
7) Other methods
8) Container methods.

Because many of the methods require separation of the roots from the soil, the general washing techniques are summarized in a separate chapter (Sect. 11). The various root parameters and their determinations are also described separately (Sect. 12) so that repetitions are avoided.

2.3 Selection of the Best Method

The choice of the best root-study method depends on the research aim of the investigator. There can be no ideal method for answering all questions. Often it will be advisable, sometimes also necessary, to use two methods simultaneously.

Every method has shortcomings. Although nearly every research worker who has done ecological root studies has altered existing methods or has created better techniques, most of the root-study methods are still tedious and time-consuming. Generally it can be said that the more accurate the method, the more laborious it is.

For the selection of the best study method in an individual case it is therefore not only necessary to be familiar with the existing methods, but also important to know how much time and labour will be necessary to obtain the required root data. Unfortunately indications of the time and labour needed are very seldom found in the literature; in this book in most cases also only general recommendations can be given.

3. Excavation Methods

3.1 Introduction

To expose the complete root system of a plant, one must remove the surrounding soil, which is normally accomplished by careful excavation of the soil surrounding the individual roots. Depending on the objective of the study, this procedure may be accompanied by simultaneously drawing or photographing the position and relationship of the various roots exposed. This traditional excavation method, also referred to as the skeleton method, is the oldest method used in ecological root research (Hales, 1727; Du Hamel Du Monceau, 1764/65; Thiel, 1870; Hays, 1889; Kraus, 1892–1896; Fruhwirth, 1895; Schultz-Lupitz, 1895; Ballantyne, 1916, and others). J.E. Weaver, the leading root ecologist during the first half of the twentieth century (see Sect. 2), developed the excavation technique into a well-founded, scientific procedure. Although his techniques have been refined by a number of workers, the basic principles outlined by Weaver (1926) are still valid today.

3.2 Outline of the Classical Method

3.2.1 Selection of the Plant

The plant selected for excavation should generally be a normally developed plant surrounded by others of its kind. Due to the lack of competition, isolated plants can vary considerably in top and root development from those surrounded by others.

After the plant has been chosen, its height, development stage, and special morphological features should be determined and recorded. Before starting excavation, the top parts of the plant must be removed or secured and the root crown firmly secured to maintain it in its original position. When root systems of large trees are to be exposed, it is usually wise to remove the stem and crown for safety reasons.

3.2.2 Digging the Trench

After the specimen has been selected, a trench is dug at a distance from the plant. This trench should be far enough from the plant stem to ensure that no laterally growing roots are destroyed, yet close enough to minimize the amount of soil that must be excavated. For grasses or herbs, 20–80 cm distance will generally be appropriate, depending on the species and growing stage. Large trees may have lateral roots extending 10 m or more from the stem, and the trench must be located appropriately. If the probable horizontal extension of the root system to be excavated is unknown, it is advisable to remove the surface layer of the soil carefully, beginning at the stem, to determine the full horizontal extent of the main lateral roots on the side of the plant where the trench is to be dug.

In Northern latitudes the trench should be dug on the north side of the plant, so that the excavated roots are not exposed to direct sunlight. Kutschera (1960) proposed digging the trench on the south side of the plant so that the investigator shades the exposed roots with his body while preventing the sun's rays from shining into his eyes.

The trench should be at least 1 m wide to facilitate working. Its depth should extend 20–30 cm below the deepest roots. Spade and shovel are the main tools to dig the trench. It is recommended that the excavation site be kept as free as possible of spoil material, to facilitate the selection and excavation of another plant in the same trench in case the planned excavation of the selected plant should fail.

The width and depth of the trench varies not only with the plant species but also with soil type. In very sandy soils, the trench should be fairly wide in order to minimize the danger of the walls caving in on to the worker. To reduce the safety hazard to those working in the trench, the walls of deep trenches in sandy or non-cohesive soils should be braced or shored up. The caving-in problem is much less in loessial or cohesive soils and small steps can be dug.

Where feasible, a trencher or backhoe may be utilized to reduce the labour involved in preparing the initial trench. It must be realized that mechanized trenchers can only be utilized in opening up the work area, and cannot be used actually to expose the roots themselves, due to the intensive nature of the equipment and the resulting damage to the root system. Cullinan (1921) used dynamite in root studies of apple trees; but in general this technique cannot be recommended because many of the finer roots and some of the larger roots may be broken, and a distorted picture of the root system may ensue.

The initial trench should be dug with nearly vertical walls. If the maximum rooting depth is known or expected to be shallow, the trench should be dug in one step. For deep-growing plants, it can be advisable to dig the trench in steps, so that digging alternates with excavating the roots. This procedure can save unnecessary digging and divides the heavy work into short intervals (Kutschera, 1960). In such cases it can be helpful to construct the trench with different levels, so that the soil can be removed from one level to another until it is thrown out of the trench on to the soil surface.

3.2.3 Excavating the Root System

The next step is to expose the root system. One should begin by carefully removing the soil from the plant side of the trench, starting with the surface soil, and gradually working downward and into the face of the exposed profile (Fig. 3.1). The soil must be removed particle by particle to avoid root destruction. Wherever feasible, the soil particles should be removed in a direction parallel to the roots, as

Fig. 3.1. Starting dry excavation of root systems by means of needles and similar tools

roots will resist a much greater pull if the applied stress is parallel to the direction of their growth.

It should be emphasized that it is not good practice to expose a root only until it becomes small in diameter and then to estimate the remainder of its length (Weaver, 1926). Often such roots may increase in size, branch one or more times and grow to considerably greater lengths. The only sure method to determine the exact length of a root is to find its tip.

The careful removal of soil particles from around the root system requires a wide variety of small hand tools. The traditional excavation instrument used by Weaver (1926) was a small hand pick with a cutting edge on one end and a long narrow tapered point on the other. Other workers have used ice picks, small metal forks, screw drivers, forceps, spatulas, small dental picks and sharp pointed needles of different sizes. The choice of the correct instrument depends on the type of root. If large woody tree roots are to be excavated, one would use the larger, coarser instruments such as ice picks, metal forks, and screw drivers. For plants with finer root systems, such as grasses, fine and delicate picks, sharpened needles, and camel hair brushes should be used.

Fig. 3.2. View of a raspberry root system showing labelled and marked roots for better drawing in the course of a dry excavation (Christensen, 1947). With kind permission of the East Malling Research Station, England

The consistency of the soil has a major influence on the success of the excavation method. Roots are much easier to expose in friable, well-structured, or single-grain soils than in dense, hard soils. Soils with high clay content when dry can be so hard that it is impossible to remove the soil particles without breaking many roots (Tharp and Muller, 1940). Then it is advisable to moisten the soil with water overnight, which will greatly facilitate excavation. But the soil should not be too wet, because then finer roots may be broken off and lost during the process of removing the soil (Tornau and Stölting, 1944).

3.2.4 Drawings and Photographs

Simultaneously with the excavation procedure, careful measurements and drawings should be made from the root system. The reduction scale used depends on the original size of the root system. For annual field crops Weaver (1926) recommended a reduction of two-thirds which permits drawing even the finest roots. Where the root system exceeds 1 m in depth greater reduction will be necessary.

One should always check the position and length of the exposed roots before beginning the drawing. Sometimes roots must be bent back out of the way, labelled

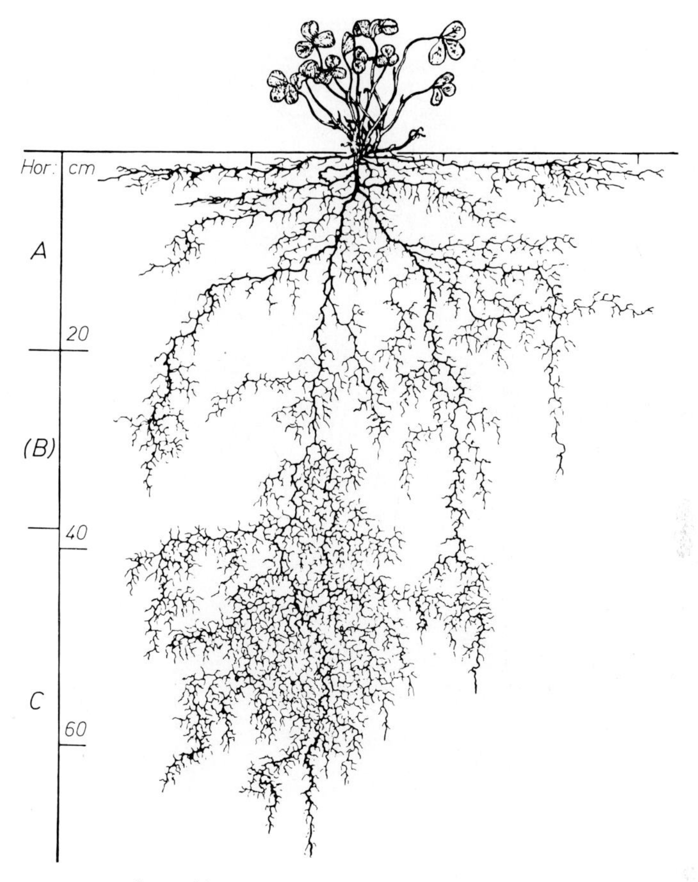

Fig. 3.3. Root system of a white clover plant (*Trifolium repens* L.) drawn during dry excavation (Kutschera, 1960)

or detached (Fig. 3.2). Kutschera (1960) proposed using thin transparent templates with a marked grid system for locating and measuring the roots. Aubertin (1971) used a three-dimensional grid upon which he drew the size, shape, and special relationship of all the roots found in each 15 × 15 × 15 cm cube of soil.

Generally the excavated root system is drawn in a vertical plane. This means that all roots should be placed and drawn in their natural position with reference to the soil surface with a vertical line from the base of the plant stem. Different root diameters can be indicated through the use of pencils of different hardness or colour.

In many cases the number of branched roots in a given soil layer may be too large for all to be drawn on the graph paper at the selected scale. It is then recommended that the major roots be drawn on the original scale and a supplementary drawing made on a larger scale. The original root drawings can be transferred or copied, then photographically reduced or enlarged to the desired

Fig. 3.4. End of a single maize root (*Zea mays* L.) showing the position where branching is starting

size. Such drawings show very impressively the extent of the total root system of a plant (Fig. 3.3). These root drawings can be more instructive than a photograph.

In situ photographs made of single roots or special parts of the root system during excavation may be quite valuable in interpreting the data obtained, for example the thickening of root tips in a compacted soil layer or the beginning of branching of a root (Fig. 3.4). General hints for photographing excavated roots at natural sites are given by Glatzel (1964). As roots are frequently dark in colour, they generally do not contrast very well with the surrounding soil and are hard to distinguish in a photograph. To alleviate this problem, the roots may be painted or sprayed with bright paint. White or yellow enamel or aluminium paint have proved satisfactory (Haas and Rogler, 1953; Schultz and Biswell, 1955; Parmeijer, 1956; Bilan, 1960). With modern photographic flash units this is now less of a problem.

3.2.5 Preparation and Storage of Excavated Root Systems

Excavated root systems from non-woody plants may be pressed between blotting paper, like leaves or other plant parts. Before pressing, the roots should be

Fig. 3.5. Root system of a 16-years-old Cox's Orange apple tree grown on a brick earth soil showing main root system near surface but many roots growing below 1 m. (With kind permission of the East Malling Research Station, England)

arranged, so far as feasible, into their natural growing position according to the drawings made in the field. When dry, the root systems may be assembled into Riker mounts or fastened with transparent tape onto poster board or sheets of cardboard and stored in a dry room. With time the preserved root systems of most plant species will become brown and brittle. As a result, colour studies should be made on fresh roots.

Roots may also be stored in alcohol or in other organic liquids (see Sect. 11.3). This method tends to preserve the roots' natural colour and prevents them from becoming dry and brittle. Such preservation and storage is essential if the roots are to undergo subsequent cytological research. Preserving excavated roots by spraying with translucent synthetic products has also been done, but not always with success (see Sect. 4.6.5). The drying and mounting technique is still the most economical for preparing and preserving excavated root systems.

Tree root systems are generally not prepared and stored. After excavation, and when the in situ measurements and drawings have been made, the root system is usually transported to a more favourable location for photographing. By putting crown and root system together very impressive photographs can be taken (Fig. 3.5). Then the root system is normally discarded, although selected portions may be preserved for further studies.

3.2.6 Review of Applications

The dry excavation method has been used by a large number of research workers during the past fifty years. The following selected references are representative of the large volume of root-excavation literature and are included to provide a brief synopsis of the many applications to which root excavations have been put.

Weaver and his co-workers belong to the vanguard of root excavators. They concentrated their efforts mainly on prairie plants and annual cultivated crops (Weaver, 1915, 1919, 1920, 1925, 1926, 1930, 1958a, 1958b, 1961; Weaver and Christ, 1922; Weaver et al., 1922; Jean and Weaver, 1924; Weaver and Bruner, 1927; Weaver and Clements, 1938; Weaver and Albertson, 1943). Excavations of root systems of different grasses and herbaceous plants were made by Metsävainio (1931), Sperry (1935), Linkola and Tiirikka (1936), Holch et al. (1941), Keil (1941), Steubing (1949), Kullmann (1957a, 1957b), Coupland and Johnson (1965), and Lichtenegger (1976). The roots of sand dune plants were investigated by Dittmer (1959a). Desert plant roots were studied by Kausch (1959) and Ludwig (1977). The root systems of various agricultural plants were excavated and studied by Bruner (1932), Worzella (1932), Farris (1934), Kiesselbach and Weihing (1935), Weihing (1935), Gier (1940), Leonard (1945), Zopf and Nettles (1955), and Gwynne (1962).

The excavation method was used extensively at the East Malling Research Station, England, to study the root systems of fruit trees, primarily apple trees (Rogers and Vyvyan, 1928, 1934; Rogers, 1933, 1934, 1939c; Beard, 1943; Christensen, 1947; Coker, 1958b, 1959; Atkinson et al., 1976). Other studies which involved the excavation of root systems of orchard trees were undertaken by Kvarazkhelia (1931), Yocum (1937), Kemmer (1956, 1963b, 1963c, 1964), and Kolesnikov (1971).

Root systems of forest trees have also been studied by a number of researchers. The exemplary tree-root study of Weaver and Kramer (1932) must be mentioned, as also the investigations of Hilf (1927), Vater (1927), Lemke (1956), Melzer (1962a), and Köstler et al. (1968) in Germany, the extensive studies of Laitakari (1929a, 1935) in Finland, and Yeatman's (1955) studies in England. Additional examples of tree-root excavations are to be found in the proceedings of the First International Root Symposium held in Leningrad (Keresztesi, 1968; Zapryagaeva, 1968).

The most extensive and comprehensive root excavations since the classical work of Weaver has been done by Lore Kutschera in Austria. In her studies she has investigated nearly all non-woody species, grasses, herbaceous plants, and agricultural crops in Middle Europe. Parts of her findings have been reported in the comprehensive root atlas (Kutschera, 1960) the first of several planned volumes.

3.2.7 Advantages and Disadvantages

The excavation method provides a clear picture of the entire root system of a plant as it exists naturally. The length, size, shape, colour, distribution, and other characteristics of the individual roots making up the root system may be studied directly. In addition the interrelationship between competing root systems of other plants can be observed (Coker, 1959; Kolesnikov, 1971).

By use of this method the phenomenon of root grafting can be studied easily (Graham and Bormann, 1966). The excavation method is also quite useful in evaluating tracer techniques (Staebler and Rediske, 1958; Bormann and Graham, 1959), and should be used more frequently by grafting and tracer investigators to develop a fuller understanding of the grafting phenomenon and the limitations of tracer techniques.

The excavation method is unsurpassed for the in situ study of morphological changes occurring in the roots. Through this method it is often possible to relate these changes to the surrounding soil. Abnormal features of single roots may provide important clues to poor soil conditions (Taubenhaus and Ezekiel, 1931; Taubenhaus et al., 1931; Kemmer, 1962; Taylor and Burnett, 1964).

No other method can substitute for excavation in studying the growth of roots in soil cracks and biological channels. The excavation method is especially well adapted for studying roots growing in earthworm channels, as attested by the many German scientists who have worked in this field (Thiel, 1870; Hensen, 1892; Goethe, 1895; Böhme, 1925, 1926/27; Köhnlein, 1955, 1960; Köhnlein and Bergt, 1971; Böhm, 1974c). The same can be said for investigating roots growing in old root channels (Gaiser, 1952; Kemmer, 1963a; Aubertin, 1971).

The excavated root systems can also be converted into quantitative data (Atkinson et al., 1976). The information on the drawings can also be quantified. In tree root studies the frequency of large roots can only be quantified exactly by total or partial excavation.

On very stony or in dry sandy soils the excavation method will often be the only effective means which can be used to study the root systems (Petrov, 1968). This is also valid if the asymmetric development of tree-root systems growing on steep slopes is to be studied (Kvarazkhelia, 1931; Hellmers et al., 1955; Zapryagaeva, 1968).

The excavation method requires a large amount of physical labour and is very time-consuming. This may be acceptable when studying annual plants or young trees (Finney and Knight, 1973), but can be unacceptable and often be in no relation to the real information gained in the study of the root systems of large trees. According to Preston (1942) one man needed five weeks to excavate, measure, and record the root system of a 15-year-old lodgepole tree. Rogers and Booth (1960) reported that nearly 60 tons of soil had to be moved in the excavation of a mature orchard tree.

More than one specimen is needed to be sure that the root-system information obtained is representative of the plant species growing on the given site. Thus, whenever feasible, it is advisable to dig out more than one plant at each site. However, each additional plant multiplies the time and labour needed, even though some time and labour may be saved by using the same trench in the excavation of different plants. In areas where manual labour is cheap the time-consuming excavation procedure may be not a disadvantage. To minimize the number of plant species to be excavated, some investigators frequently use simultaneously the excavation method with other root-study methods more adapted to statistical calculations (Rogers and Vyvyan, 1928, 1934; Jacques, 1941; Salonen, 1949; Bul'Botko, 1973).

The woody roots of trees and shrubs are generally stronger and more resistant to breaking than are the fibrous roots of grasses or annual plants. Therefore it can be said in general that the excavation method is more suitable for studies involving trees and shrubs than for grasses or annual crops.

3.3 Modifications of the Classical Method

3.3.1 Excavations with Water Pressure

One of the earliest modifications was to utilize water under pressure to remove soil particles from the root system. Often referred to as the wet excavation procedure, this method, in contrast to the dry excavation method of removing the soil with scrapers and similar tools, utilizes water pressure to loosen and flush away the soil particles. Although the wet excavation technique was used very early by Schubart (1855) in Germany and by several Russian research workers (Böhme, 1927/28), the fundamental methodology was not developed until Stoeckeler and Kluender (1938) utilized the technique on a large scale.

The amount of water pressure required depends on the soil conditions and to some extent on the nature of the root systems. The object is to have sufficient water pressure to dislodge and flush away the soil particles rapidly without causing breakage and loss of the fine roots. In general, two types of nozzle should be used: one which supplies a single narrow stream of water at relatively high pressure and another which produces a fine spray or mist. To improve efficiency a nozzle which supplies a pulsating jet of water at moderate low pressure may be preferred.

Elaborate equipment is not necessary. For example Hellmers et al. (1955) simply used tank trucks equiped with water pumps, while Taerum and Gwynne (1969) used water from a gravity-fed mobile 1000-l tank 50 cm above ground level. Even simple hand- or back-carried garden sprayers may be used effectively.

Before starting the wet excavation procedure, the initial trench must be dug as described in Section 3.2.2. Since water and mud will tend to collect in the bottom of the trench, it is advisable to dig the trench slightly deeper than for dry excavation. It is also recommended to slope the bottom of the trench so that the water will flow to and collect at one end of the trench, out of the immediate work area, where it may be bailed or pumped out.

As in the dry excavation method, the top parts of the plant must be removed or secured in their original position. Then, beginning at the soil surface, the plant side of the trench wall is sprayed with water from the nozzle in such a way that the soil is loosened and transported to the bottom of the trench. Initially, most of the soil can be removed with the high-pressure nozzle. However, as soon as fine roots become exposed, the fine-spray nozzle should be used and the water pressure reduced to prevent damage to them.

After some experience, the investigator quickly obtains a feeling for the most effective water pressure and for the spray pattern which will most rapidly remove the soil particles without injuring the fine roots. In general, it is advantageous to move the nozzle slowly back and forth horizontally and vertically to keep a relatively smooth and vertical plane on the working face of the profile wall. Overhangs should be avoided as they may break off, and damage the root system. As in the dry excavation method, the exposed roots should be drawn to scale and photographed as the excavation proceeds.

The wet excavation method has mainly been used in studying plant species with woody roots such as trees and shrubs (Kolesnikov, 1930; Nutman, 1933a, 1933b; Day, 1944; Upchurch, 1951; Hellmers et al., 1955; Stout, 1956; Seshadri et al., 1958; McMinn, 1963; Singer and Hutnick, 1965; Kummerow et al., 1977; Hoffmann and Kummerow, 1978) although with adequate skill and patience, cereals and other grasses may be wet-excavated with success (Böhme, 1927/28; Bose and Dixit, 1931; Taerum and Gwynne, 1969).

In sandy soils wet excavation works much faster than dry excavation, but with increasing clay content, the advantage in time decreases (Stoekeler and Kluender, 1938). Theoretically, filling the entire excavation trench with water and allowing it to stand for several days should soften the finer-textured clay soils, and make it easier to wash the clay from the roots. Unfortunately, this good theory does not hold up in practice (Hellriegel, 1883; van Breda, 1937).

The main disadvantage of wet excavation is that the finer roots, when wet, tend to cling together, thus preventing the effective study of root branching or the relationship of the root to specific soil features. Another drawback is that the amount of water necessary produces large quantities of mud. On the other hand, wet excavation has the advantage over dry excavation, that with low water pressure and suitable soil conditions the breakage of the fine roots is much less. Therefore, depending on the plant species, soil conditions, and research objectives, one method may be superior to the other, or a combination of wet and dry excavation may prove to be the most efficient and economical procedure (Seshadri et al., 1958; Magnaye, 1969).

3.3.2 Excavations with Air Pressure

The removal of the soil particles from around the root system may also be accomplished by the movement of air in one of two ways—either by being blown away by air under pressure or by being sucked up by air rushing into a vacuum.

The use of compressed air to blow away the soil particles was recommended by van Breda (1937) who investigated several plant species in South Africa. He uncovered the larger roots of the plants by means of wire picks and other coarse tools. Where profuse fine-branched roots were encountered, he used fine picks to loosen the soil, which was then blown away by an air pump. When working with friable loam or sand, loosening the soil with a pick generally was not necessary.

According to van Breda, the compressed air technique was effective in practically all soil types, but did not work well in wet soils. An advantage is that the fine roots are not damaged as much as when using the normal dry excavation method.

The usefulness of compressed air for excavating root systems of forest trees was tested by Weir (1966). His equipment consisted of a trailer-mounted compressor rated at about 7 atm delivery pressure, a nearly 2-cm diameter pressure hose, and a

150-cm long piece of galvanized pipe at one end to serve as a nozzle. The pipe was fitted with a shut-off valve for regulating the air pressure. The technique worked successfully on a dry sandy loam interspersed with rocks of various sizes. The time required to excavate the root system of one tree was estimated to be nearly the same as would be required using the wet excavation method.

The usefulness of this technique with air pressure cannot be satisfactorily evaluated. Its greatest potential seems to be in combination with dry excavation on suitable sites. By use of this combination, it is probable that the time needed to excavate a given root system may be reduced substantially.

The vacuum excavation technique has not been used enough to provide information for evaluation. Aubertin (1971) used the technique in combination with dry excavation to dissect a large block of soil in a mature oak-hickory forest. He used a portable, heavy-duty, industrial vacuum cleaner equipped with a flexible duct and a nozzle with an opening of about 2.5 cm. Electrical power was supplied by a portable gas-fuelled alternator which generated the current. In operation, the nozzle is placed a few centimeters from the work surface and the soil particles, detached through the use of small sharpened probes or picks, are immediately sucked into the nozzle, leaving the working area continually free of loose soil. Examination of the contents of the vacuum cleaner and the root system itself revealed that only very few of the fine and none of the coarse roots were broken off and vacuumed up.

One big advantage of the vacuum technique is that the roots do not tend to cling together as they do by wet excavation. The technique works best in friable moderately dry soils, but it may be used effectively also in other kinds of soil.

3.3.3 Excavations in a Horizontal Plane

Excavations in a horizontal plane without digging a trench are to be preferred for those plant species having primarily horizontally growing root systems. This technique was first successfully demonstrated by Cannon (1911), who dug up the root systems of numerous desert plants. Later it was used by many investigators for other non-woody plant species (Markle, 1917; Weaver and Bruner, 1927; Linkola and Tiirikka, 1936; Kutschera, 1960; Dittmer and Talley, 1964, and others). However, this technique was often used in combination with the technique of excavation in a vertical plane.

The main application of root excavation in a horizontal plane is to be found in the study of tree roots. In use, a circle with a maximum diameter of about one and a half times that of the horizontal projection of the branch spread is marked around the tree selected for investigation (Kolesnikov, 1971). The excavation starts from the trunk, which should be secured or removed for safety reasons, with the gradual removal of the soil layer by layer until the first main roots are exposed. Normal excavation tools should be used to avoid damage to the roots (see Sect. 3.2.3). Kolesnikov (1971) recommended the use of special two- or three-pin metal forks and chisels. Use of a heavy-duty industrial vacuum cleaner may save substantial time and effort and improve the efficiency of the operation (see Sect. 3.3.2).

Fig. 3.6. Root systems of forest trees exposed by removal of the forest floor and the upper centimeters of soil. The white rope squares are 61 cm on a side. (Lyford and Wilson, 1964)

After the first layer of horizontally growing roots is exposed, a grid net made of rope or wire is stretched over the soil surface (Fig. 3.6). A suitable size for the grids is 50 × 50 or 100 × 100 cm. Such a grid system makes the work much easier and enhances the accuracy of the drawing (Breviglieri, 1953; Bargioni, 1959a, 1959b, McMinn, 1963; Lyford and Wilson, 1964). A metal or wooden platform should be installed directly above the grid net to provide easy access to every part of the root system.

In most tree root excavations, one generally looks at only the thicker roots in order to obtain an overall conception of the gross morphology of the root system of the tree. Due to the tremendous amount of labour involved, a detailed mapping of the fine roots (less than 2 mm) generally cannot be made for larger trees. In order to determine the amount and extent of these fine roots, other methods such as sampling or profile wall methods should be used.

Often the total root system is not excavated down to the deepest growing roots, but only the main horizontally growing roots in the upper soil layers are studied and mapped in a horizontal plane (Fig. 3.7). Depending on the study objective, it may be highly informative also to draw roots from other nearby trees and indicate the location of their stems on the drawing (Kolesnikov, 1971). The use of different colours will enhance the visualization of competing root systems. One drawback is that generally the excavation in the horizontal plane will result in more root damage and loss than by working in the vertical plane.

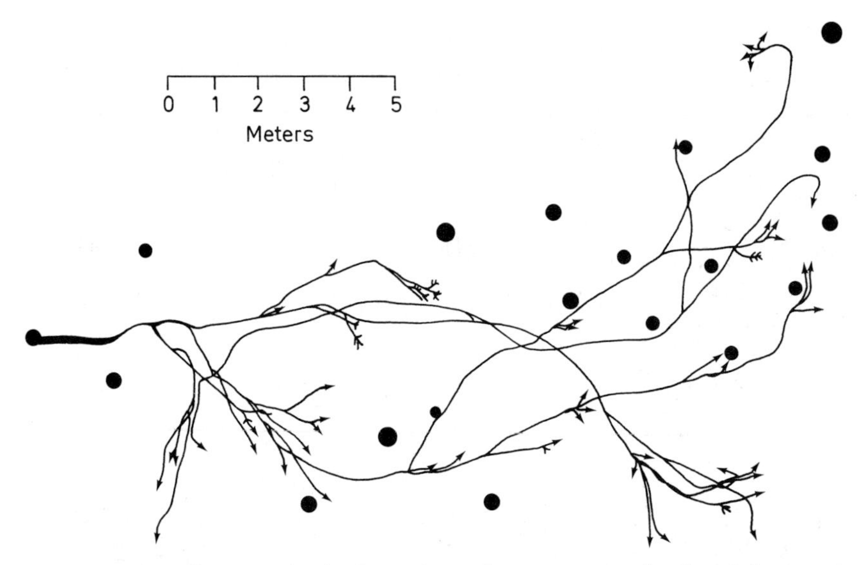

Fig. 3.7. Plan view diagram of a horizontal woody root system developed from a single lateral root of a red maple tree about 60 years old. *Solid circles* show the location of other trees in the stand. *Arrows* indicate that root tips were not exposed and therefore these roots continue farther than is shown (Lyford and Wilson, 1964)

3.3.4 Sector Method

The time- and labour-consuming procedures for excavating the total root systems of plants, especially of trees, led to modified, more economical techniques. It soon became obvious that exposing only a small representative portion of a given root system could give, in many cases, sufficient and adequate information on the entire root system. Rogers (1932), Nutman (1933a), and Krauss et al. (1934) utilized the sector method in their numerous tree root investigations. In so doing, they excavated only a sector of the total root-containing area surrounding each tree. Later this technique was used in forest tree studies (Joachim, 1953; Schoch, 1964; Köstler et al., 1968; Jahn et al., 1971).

The most common procedure for carrying out a sector excavation consists of digging a hole 1 or 3 m long by 1 m wide, approximately 50 cm from the trunk and perpendicular to a radius (Köstler et al., 1968). The depth of the hole will vary with size and depth of the root system. The soil between the hole and the trunk is then excavated and the heart of the root system is exposed. The exposed portion of the root system is then drawn or photographed (Fig. 3.8). Jahn et al. (1971) used a stereo-camera to obtain photographs of the exposed root systems and obtained fascinating pictures.

It may not always be necessary to expose a cross-section through the heart of the root system. According to Nutman (1933a), who worked with coffee trees, it is often sufficient to expose only one side of the root system far away from the stem. This view is supported by Russian scientists who believe that an adequate characterization of the total root system of a tree can be obtained by unearthing

Fig. 3.8. Root system of a 50-year-old beech tree exposed by using the sector method. (Köstler et al., 1968)

only a small portion of it (Kolesnikov, 1971). However, to avoid possible misinterpretation of the data due to asymmetrical root distribution, it is advisable to utilize several sector excavations located at random distances within the expected area of root extension and at random directions from the stem.

Exploratory root investigations were made by Karizumi (1957) in Japan using the sector method. In addition to the normal sector excavation for obtaining an indication of the gross morphology of the root system, he counted the number of roots more than 2 mm in diameter and estimated the roots below this diameter by frequency.

Although the sector method, as outlined above, is generally more adapted for tree-root systems, it may be used also with advantage in the study of non-woody plant-root systems. Gliemeroth (1953b) and Frese et al. (1955) successfully used this method to obtain excellent pictures showing the influence of manuring on the root system of sugar beets. In general, however, the sector method is not suitable for the study of plant species which characteristically have many fine roots, as these roots normally do not remain in their natural position after excavation.

4. Monolith Methods

4.1 Introduction

These methods require the taking of soil monoliths and separating the soil from the roots by washing. This can be done directly in the field, but generally the soil samples are transported to a special washing place. These methods are frequently applicable for investigation where the aim is to make a quantitative determination of the roots, but in many cases this is supplemented with visual pictures.

Small monoliths can be taken with simple tools, such as spades, but with increasing size of the monoliths, power-driven mechanized machines must be used. In most cases trenches must be made before starting the sampling procedure. Monolith methods are mainly used by researchers in agriculture.

4.2 Simple Spade Methods

To obtain rough information about the roots in the upper soil horizons, small monoliths about 20 cm in square can be taken with a spade (Görbing, 1948). These soil blocks are gently shaken by grasping the aboveground parts of the plants with the hand and the exposed part of the root system can be judged. With this simple technique Pittman (1962) examined the growth direction of young roots of winter wheat.

Generally it is more common to put the monoliths dug out by a spade in small boxes with a sieve bottom, and after soaking, to wash the root systems free with a hose or spray. Then the root systems can be classified visually by size (Hamilton, 1951). Such simple visual classifications were also made for testing herbicide damage on cereal roots (Maas, 1968, 1970). Classification scales can also be made for plant species with tap roots where the extent of the development of lateral roots is a distinguishing mark (Klebesadel, 1964).

Such spade methods, with additional visual estimations of the exposed root systems, are relatively quick and are adequate for solving specific questions. In many cases they are used only as a first preliminary test before starting with a more accurate, quantitative study method on the same site. Instead of by spade, the monoliths can also be taken by means of special cutter blades which are driven into the soil by a hand-operated impact driver (Rivers and Faubion, 1963).

Fig. 4.1. Taking soil-root monoliths starting from a trench

4.3 The Common Monolith Method

4.3.1 Square Monoliths

A proven technique consists of digging a trench about 1 m long to the maximum rooting depth, and then taking away monoliths from the side wall layer by layer. The height of the monolith sample depends on how accurately the root distribution in a soil profile is to be determined. Taking samples from 10-cm layers is common. From one 10-cm layer from a profile wall of 1 m length, about five or more subsamples can be taken.

The size of such soil monoliths varies from small soil prisms of about $10 \times 10 \times 10$ cm to large blocks of 1 t in weight and more dependent on the plant species and the aim of the research. Generally the volume of a common monolith varies between 1000 and 5000 cm^3 (Burton, 1943; Klapp, 1943; Paavilainen, 1966; Safford, 1974; Ehlers, 1975; Köpke, 1979).

Before starting the removal of the monoliths from the soil, the profile wall must be flattened, using a plumb line. All loose soil should be removed from the bottom of the trench. Then the size of the monoliths to be taken is marked on the profile wall with a needle.

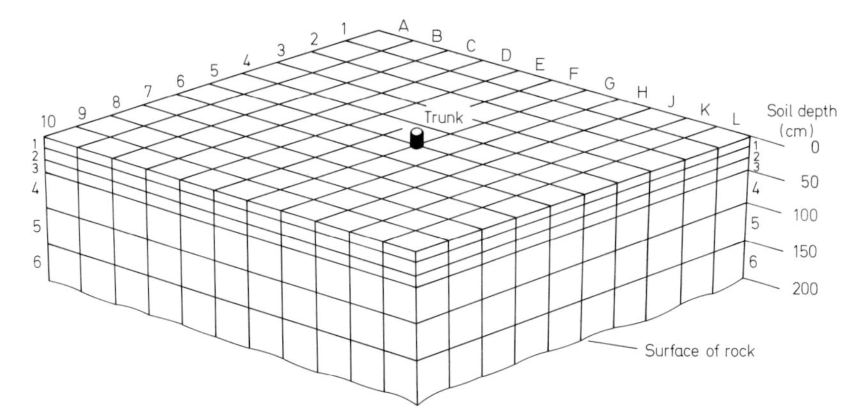

Fig. 4.2. Diagram showing the arrangement for taking soil-root monoliths of a single tree. (Isometric projection after Rogers and Vyvyan, 1928)

To extract the soil monoliths broad knives and metal sheets sharpened at one side which can be beaten into the soil with a hammer are used (Fig. 4.1). This beating will in most cases be necessary when the soil is dry. It is advisable to place a foil on the bottom of the trench, so that crumbling soil lost from the monoliths can be collected easily. The complete soil sample is put into a pail and the roots are separated by washing (see Sect. 11).

Especially in root investigations on agricultural plants, much larger monoliths have often been taken (Land and Carreker, 1953; Balázs, 1954; Bommer, 1955; Fröhlich, 1956; Bobritzkaja, 1960; Asana and Singh, 1967; Lieth, 1968). In some cases the size of one single monolith was up to 100 × 100 × 50 cm (Günther, 1951; Könnenecke, 1951). Only spades and shovels are used for taking such large monoliths. The soil is thrown into adequate containers and carried to the washing place. If the plants grow in rows, the monoliths are prepared in such a way that the plants are always standing in the middle of the monolith. Replications in experiments of such large monoliths have seldom been done.

For quantitative studies of the whole root systems of trees which occupy large areas of soils, immense amounts of soil must be moved. Rogers and Vyvyan (1928, 1934) studied the root system of apple trees with such a giant soil block technique, and moved as much as 60 tons of soil per tree. They also dug a trench across the area possibly occupied by the roots of the tree, and removed the soil in blocks of known size and position (Fig. 4.2). In a similar way, such more or less giant operations were used by Lee (1926) and Lee and Bissinger (1928) with sugar cane, by Guiscafré-Arrillaga and Gómez (1938, 1940, 1942) with coffee trees, by Coker (1959) with apple trees, and by Bilan (1960) with pine trees.

The aim of obtaining only roots from the tree selected for investigation will seldom be reached with this technique. Roots from neighbouring trees will also be growing in the area of the excavated soil block. On the other hand, roots from the tree under study will also grow beyond the excavation limits. Sometimes this is not a major problem. In fields or orchards where plants are growing at uniform distances, the loss of roots beyond the limits of the excavation place is generally

Fig. 4.3. Free-standing soil monolith prepared by means of a spade for determination of roots in upper soil horizons

compensated by the roots of the neighbouring rows which extend into the excavation region (Guiscafré-Arrillaga and Gómez, 1938). Nevertheless, at present the use of such giant monolith techniques to obtain quantitative data on the whole root system of a single plant can only be recommended for comparative ecological studies if adequate manual labour is available.

If only roots in the upper soil layers are to be studied, the technique of preparing free-standing monoliths without digging trenches can be used (Fig. 4.3). After the monoliths have been prepared with a spade they are cut at the bottom and transferred on sieve boxes to a washing place. This technique has been widely used by Köhnlein and Vetter (1953) for investigating the total root mass of agricultural crops in the plough layer of the soil.

Instead of preparing free-standing monoliths, special soil samplers can also be used for soil monoliths taken from the upper soil layers. Numerous research workers have used different kinds of frames and metal boxes, predominantly for root studies of agricultural crops and forest trees (Jacques, 1937a; Lamba et al., 1949; Bohne and Garvert, 1951; Könekamp, 1953; Bohne and Greiffenberg, 1954; von Hösslin, 1954; Heikurainen, 1955a, 1955b; Könekamp and Zimmer, 1955; Kmoch et al., 1958; Long, 1959; Roder, 1959; Stucker and Frey, 1960; Schulze and Mues, 1961; Foth, 1962; Zöttl, 1964; Walker et al., 1976).

The size of the common sampler types ranges from 5 × 5 cm to 30 × 30 cm in square and 10 to 50 cm in length. Usually the sampler is an open three-sided steel box. The thickness of the steel should be at least 6 mm. One side is removed so that the soil will not be compacted when the box is forced into the soil. The other three sides are sharpened. The top of the box is reinforced with an additional steel plate. The

Fig. 4.4. Driving a steel box into the soil for obtaining a soil-root monolith

Fig. 4.5. Sectioning the soil-root monolith in the steel box into subsamples by means of a knife

box is driven into the soil with a heavy hammer. To protect the upper steel plate a piece of hard wood can be placed on the top during the beating procedure (Fig. 4.4).

To remove the sample some soil is dug out with a spade from the area around the outside of the sampler. Then the sampler is brought out of the hole in such a way that some soil extends out beyond the ends of the open side of the sampler. The

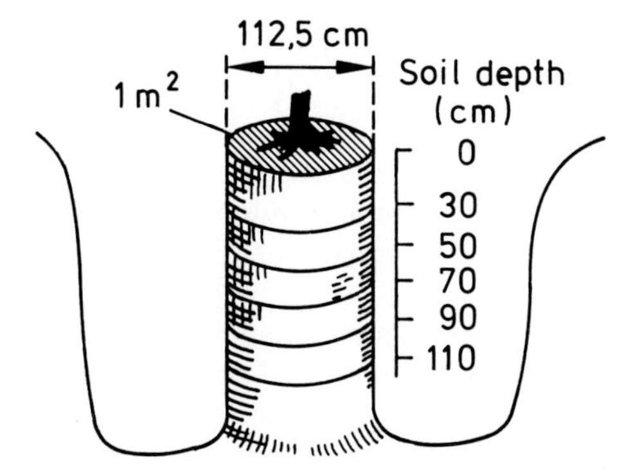

Fig. 4.6. Arrangement of taking round monoliths in tree root studies. Technique of Kreutzer (1961, 1968)

sampler is then laid down flat and the soil on the open side of the sampler is smoothed with a knife or a sharpened metal sheet. Finally, the whole soil monolith can be cut in subsamples (Fig. 4.5).

With this technique an almost exact volume of soil can be sampled, and the method is recommended when the roots in only the upper soil layers are to be studied. The sampler must always be driven vertically into the soil. In wet soils compaction can occur so that the real soil depth can differ from the depth measured with the sampler. However in taking samples only from the upper soil horizons (0–30 cm), such possible compaction can generally be ignored.

4.3.2 Round Monoliths

For root studies on trees, specifically to examine their central, vertically growing roots, the preparation of large, round monoliths has been used. Kreutzer (1961, 1968) proposed, after felling the tree, digging a circular trench around the stem, so that a cylindrical soil monolith remained under the sample tree (Fig. 4.6). To standardize such a cylindrical block method the monolith should always have a diameter of 112.5 cm, i.e., an area of 1 m^2. Depending upon how precisely the spatial root distribution is to be determined, the monolith can be levelled layer by layer with the help of knives and saws. Satisfactory results with this method will be obtained on sites where the trees are more or less regularly spaced (Karizumi, 1968).

To obtain round monoliths from soils planted with non-woody plants, large iron cylinders up to 40 cm in diameter and 100 cm in length have been used (King, 1892; Böhme, 1925; Laird, 1930). They were driven into the soil with a hammer and dug out with a spade or pulled out by means of a hook. To extract the soil from such cylinders or obtain subsamples from distinct soil layers is, however, difficult, and often impossible. Better adapted are small steel rings or small cylinders about 5 to 30 cm in diameter and up to 30 cm in length which are mainly used for obtaining soil samples from the upper soil layers (Schwarz, 1934; Heyl, 1942; Kmoch, 1952; Gliemeroth, 1953a; Wetzel, 1957/58; Malicki, 1968). These techniques have usually been replaced by the various auger techniques (see Sect. 5).

4.4 Box Methods

In contrast to the common monolith methods, the box methods described here consist of preparing large soil monoliths encased in boxes where the root system is washed free without dividing the monoliths into sections. In this way the complete or the characteristic main part of the root system of one or more plants can be obtained.

Classical studies of this technique were first successfully demonstrated by Pavlychenko (1937a) with weeds and annual field crops. His method, originally termed the "soil-block washing method" consists of isolating a square monolith of soil around one plant, large enough to contain its complete root system. For 40-day-old cereal plants Pavlychenko (1937b) used monoliths $65 \times 65 \times 105$ cm. For mature plants $80 \times 80 \times 165$ cm monoliths were used. The monolith is enclosed tightly by means of a strong wooden frame, tipped over on its side and lifted out of the pit by means of a derrick. After transporting the box to a washing place, one side of the wooden frame is removed. The monolith is soaked with water for several hours or days in a large tank which is placed at an angle of about 10° and the root system is carefully washed free (see Sect. 4.6.4).

When the root system is freed from the soil it is placed in a small tank with sufficient water for the roots to be submerged entirely when spread out horizontally along the bottom. In this position the different root parameters are counted and measured. If desired the whole root system can be mounted and photographed (see Sect. 4.6.5).

Although detailed quantitative study of the complete root system is possible with this method, only a few research workers have used it (Gier, 1940; Pavlychenko, 1942; Cook, 1943), because generally the input in time and labour is greater than with the dry excavation method. However, with no other method is the loss of the fine roots and even root hairs so minimal, so the few results obtained by this method are highly recognized.

Summarizing, the examination of a complete root system in this way is only economical for very young plants whose monolith can be moved easily by hand without special mechanical equipment.

A similar suitable box method, where only a part of the root system from one or more plants is taken, was introduced into ecological root research by Weaver and Darland (1949a and 1949b). The monolith prepared has, however, relatively small dimensions so it can be removed easily from the trench, and transported without special mechanical equipment.

After the site with normally developed plants has been chosen a trench is dug nearly 30 cm below maximum rooting depth. For only one monolith, a convenient size for the trench is 100 cm wide and 120 cm long; the side walls of the trench are nearly vertical (see Sect. 6.2.2). Then an open wooden box about 20 cm wide, 10 cm deep and 100–150 cm long is placed against the wall of the trench, and the outline of the soil monolith to be taken is marked with a pointed needle at the profile wall. The box is then removed and the soil on the sides and below the marked lines is worked away by means of a spade and knives until the whole monolith is protruding from the wall and the size is nearly equal to the inner diameter of the

Fig. 4.7. Box method of Weaver-Darland. Prepared soil-root monolith still hanging on the profile wall with plants in the middle

wooden box (Fig. 4.7). Then the box is fitted tightly over the monolith and its bottom and lower ends are braced to prevent collapse.

Finally the soil on the inner attached side of the monolith is also worked away. The soil must not be cut close to the top of the box. A ridge of the soil should protrude throughout the whole length. Then the braces are removed and the monolith can be lifted out of the trench. The ridge of the protruding soil mass is cut away so that an exact rectangle is obtained (Fig. 4.8).

After the monolith has been lifted out of the trench, it is submerged in an adequate tank of water, and with a technique similar to that of Pavlychenko, is soaked and washed (see Sect. 4.6.4). Roots that were cut from other plants are lost during the washing process if not collected on a sieve. After all the soil has been washed away the root system remains in a horizontal plane nearly in its natural position at the bottom of the box. The actual washing time for one monolith with a size of $10 \times 20 \times 150$ cm can vary between 2 and 10 man hours depending on the texture and structure of the soil.

The complete, washed root system is then transferred to a black mounting board and photographed. Finally it can be dried as a whole and preserved, or it can

Fig. 4.8. Box method of Weaver-Darland. Encased soil-root monolith ready for washing

be sectioned according to different soil depths, and the weight or other parameters determined. During the washing process, however, roots can move out of place, and if only quantitative root distribution in distinct soil depths is be investigated, it is generally more accurate to cut the monolith into sections before soaking and to wash out these segments separately.

The method of Weaver and Darland which combines qualitative, pictorial representation, with obtaining quantitative root data has been used mainly in the USA (Fox et al., 1953; Fehrenbacher and Snider, 1954; Fehrenbacher and Rust, 1956; Kmoch et al., 1957; Fehrenbacher et al., 1960, 1965, 1967, 1969, 1973; Weaver, 1961; Guernsey et al., 1969). Particularly the penetration of the root systems of field crops on different soil types was studied with great success (Fig. 4.9).

In the course of time, various modifications of the method have been described. Weaver and Voigt (1950) increased the size of the monoliths. Fox and Lipps (1955a and 1955b) replaced the back of the sampling box with a 1.2 cm hardware cloth which permitted washing from above and below. Williams and Baker (1957) used a sampling box made of iron instead of wood, whose thinner walls were easier to fit the box over the prepared soil monolith. Schulze (1971) and Frielinghaus and Spitzl (1976) used a steel frame as a box where all the sides except the bottom were made

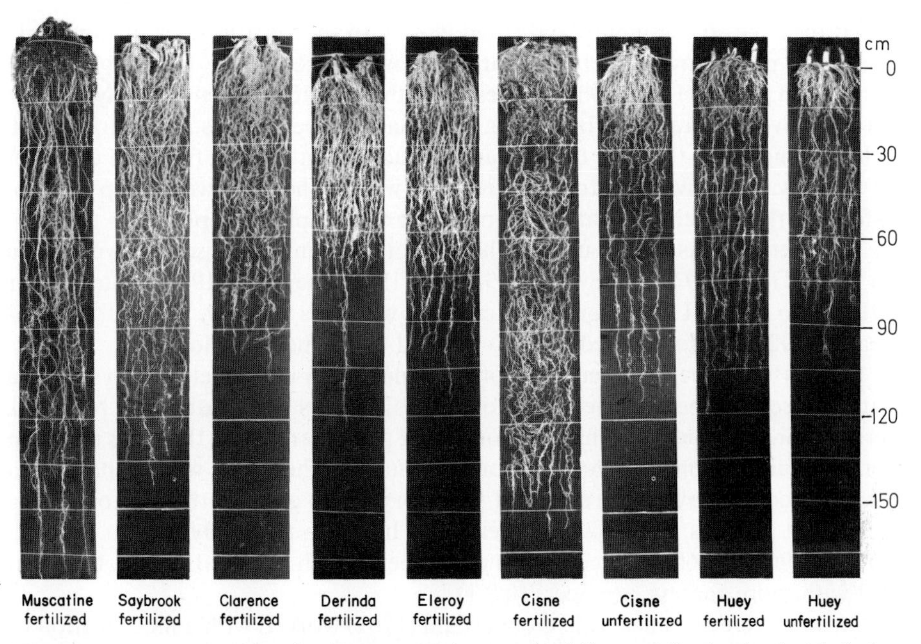

Fig. 4.9. Root systems of maize (*Zea mays* L.) grown in different soils obtained with the box method of Weaver-Darland. (Fehrenbacher et al., 1967)

of 5-mm metal mesh. Foth and Pratt (1959), mainly for demonstration purposes, recommended washing only a small layer of soil from the monolith in the box. Then the surface of the remaining soil block and the exposed part of the root system can be preserved with a vinylite resin solution.

As it is only in rare cases necessary in modern ecological research to obtain a complete root system of a single plant, the method of Weaver and Darland is a valuable alternative to the method of Pavlychenko. It is certainly still very laborious, but more effective because for the same amount of time and labour replications can be made, and therefore the ecological worth of the results will generally be greater.

4.5 Cage Methods

One of the oldest dreams of the root ecologist is to have study methods which will produce three-dimensional pictures of the root systems in their natural positions. With the excavation method such root pictures can only be obtained from plant species with strong roots, like trees. King (1892, 1893) demonstrated with a special cage-box method that three-dimensional root pictures can also be obtained from other plant species.

His method, tested with several annual agricultural crops in the field, consists of preparing a free-standing soil monolith about 60 × 30 cm in size containing one or more plants in the middle. The trenches around the monolith extend to the maximum rooting depth. Then a cage of galvanized iron or poultry wire netting is placed over the soil monolith and fitted closely to the vertical sides of the block. After this procedure numerous sharpened wires are forced through the soil monolith in parallel rows along the meshes of the netting. Next, the top layer of the soil is removed and replaced by casting in its place a block of plaster of Paris, which on hardening keeps the stem of the plant in place. Then starting from the top layer the soil is washed away in situ by a stream of water. The cross wires keep the root system nearly in their natural position, and it can be photographed.

This method was widely used at the turn of this century by several workers in the USA (Goff, 1897; Ten Eyck, 1899, 1900, 1904; Shepperd, 1905). Later Miller (1916), also in the USA, worked extensively with this method, but both he and Weaver (1926), and van Breda (1937) pointed out some objections.

Besides the immense expenditure of time and labour a complete root system of a plant can seldom be obtained. In most cases only its main parts can be saved. Furthermore, although the thicker roots of the plants remain on the wires nearly in their natural position, fine roots do not, but cling together when wet, so that only a distorted root picture can be obtained. In theory a very good method for obtaining in situ root pictures, in the last few years this has been used only seldom (Wetzel, 1957/58, 1960a, 1960b). In container experiments with artificial soils it has been used with more success (see Sect. 10.9.1).

On an extreme smaller scale Gooderham (1969) worked with a modification of such a cage method. A soil core taken out of the field (about 20 cm long by 5 cm in diameter) is transferred to a Plexiglas tube of similar internal dimensions in which holes are drilled. Then a nylon mesh is threaded through the soil core in order to keep the root system in its natural position while the soil is removed by washing. Finally, the nylon mesh is replaced by gelatin, and in this way the nearly undisturbed root system can be studied, photographed and preserved. Similar studies have been done by Bloomberg (1974). This technique is, however, primarily interesting for demonstration purposes, and cannot be recommended for routine studies in the field.

4.6 Needleboard Methods

4.6.1 General Survey

Probably the most common root-study method which combines pictorial presentation with quantitative determination of the root system of the plants is the needleboard method. By this method a soil monolith with a representative sample of the root system is taken by means of a special wooden board. Needles or nails positioned in this board keep the roots nearly in their natural position while the soil is removed by soaking and washing. The root system can be investigated and photographed as an entirety, or can be sectioned for more quantitative determinations.

The needle-, nail-, or pinboard method is a further development of King's cage method (King, 1892). Goff (1897) in the USA was the first to replace one side of King's iron cage by a wooden board with needles, and obtained very impressive root pictures from different orchard crops. Later Rotmistrov (1909) in Russia used this method in his studies of plant root systems grown in boxes with artificial soils (see Sect. 10.9.2). Maschhaupt (1915) in the Netherlands was the first to work in the field with this method for root studies of agricultural plants. During the last decade the needleboard method became popular partly due to the detailed descriptions of this technique by Schuurman and Goedewaagen (1971).

4.6.2 Construction and Preparation of the Needleboards

The dimension of the boards and the length of their needles can vary according to the size of the root systems to be sampled. To take a small soil monolith, a board size of 50 × 50 cm with needles 5 cm long is common, for larger samples sizes up to 60 × 100 cm with needles up to 20 cm can be used. Still larger dimensions are possible, but they will be governed by the mechanical equipment and manpower available to handle such large samples.

De Roo (1957) and Schuurman and Goedewaagen (1971) constructed needleboards from two similar-sized pieces of plywood, 2 cm and 1 cm thick. Holes were drilled into the thicker board in vertical and horizontal rows 5 cm apart. Then knitting needles or stainless steel wires were cut to length, sharpened at the cut ends and bent into U-shapes. These bent needles were pushed through the holes of the 2-cm board and backed up by the thinner piece of plywood. Brass screws held the two boards together and kept the needles in place when the board was driven into the soil.

Instead of U-bent wires or needles, nails or other pins can also be driven directly into the board. Bicycle spokes cut to the desired length have also been used with success.

The 5-cm distance between the needles on the board has been proved to be the most suitable. If the needles are placed closer together, the washing procedure can become difficult in the case of hard soil layers. If the needles are spaced wider apart too many roots move from their natural position and the root picture will be distorted. Only when working with rigid tree-root systems can larger distances between the needles on the board be used without difficulty.

Finally, the needleboard is painted black to provide a contrasting background for photographing the exposed root systems. To prevent rusting the needles should also be painted (De Roo, 1957). Instead of painting, black polyethylene sheeting can be stretched over the board and pressed between the needles until it is against the board (Moraghan, 1968; Taerum and Gwynne, 1969; Schuurman and Goedewaagen, 1971). This has the advantage that after washing the root system as a whole can be lifted easily from the board.

A further improvement is to cover the black painted board with a nylon net of about 5 mm mesh (Fergedal, 1967; Kirby and Rackham, 1971), or to insert a hardware cloth among the needles (Schuster, 1964; Schuster and Wasser, 1964). Net or hardware cloth can be lifted during the washing procedure to allow water and soil to pass underneath. In this manner the abrasion of roots by soil particles can be minimized.

Some research workers (Maschhaupt, 1915; Bennett and Doss, 1960; Doss et al., 1960; Raper and Barber, 1970a, 1970b; Mitchell and Russel, 1971) used only pre-drilled boards for taking monoliths. In the field, with the pre-drilled board in

Fig. 4.10. Driving a needleboard into the profile wall. The bottom plate *on the left* is driven into place before the monolith is cut free. Technique of De Roo (1957)

place, the needles were driven through the holes and inserted into the profile wall. This technique generally is not recommendable for boards of small dimensions, but it can be useful for taking giant soil monoliths (see Sect. 4.6.6).

4.6.3 Excavating the Monoliths

First a trench is dug at least 1 m long and wide. If the plants to be excavated are growing in rows the trench can be dug parallel or transversely to the rows. The distance from the face of the trench to the stem of the plant to be sampled should be determined in such a way that their crowns come to lie in the middle of the soil monolith to be held by the needles on the board after excavation. The face of the trench can be cut back to the desired distance from the plants, and then the wall is smoothed to an exact vertical plane (see Sect. 6.2.2).

After preparation of the profile wall the needleboard is placed against this vertical wall. The top row of the needles should be almost at the ground level (Fig. 4.10). Then the board is pressed into the soil by means of a jack or a hammer to the fullest extent of the needles. The enclosed soil monolith will be supported by the jack or by wooden planks.

The soil from the profile face right and left of the board and at the bottom is then sliced away with a spade and knives. Cutting the bottom free can be done with a steel plate driven into the wall (De Roo, 1957). Such a bottom plate also supports the monolith, which is important in sandy and loose soils. To minimize the danger of soil clods breaking off during the excavation procedure, a wooden or metal frame can be pushed over the sides of the monolith.

At last the inner face of the monolith is cut free. This is done either by working inwards also with a spade and knives starting from the two sides, or by driving a steel sheet into the ground from the soil surface. Such a steel sheet can also be driven into the soil before the trench is dug (Fig. 4.10), which is especially desirable in very loose soils.

The cutting free of the inner face of the monolith should not be done directly at the tips of the needles. It is advisable to prepare the monolith a little thicker because parts of the soil can break away during the cutting procedure. Finally the needleboard with the soil monolith is laid down flat, after which the excess soil from the inner face is cut away.

4.6.4 Washing Procedure

If the monolith consists of sandy or loamy soil it is directly submerged in a tank of water until the soil is saturated. Generally a soaking overnight is enough. The frame which encases the monolith should not be removed, to avoid breaking off parts of the monolith. Instead of the frame, however, a nylon net can be placed over the whole monolith, which also prevents the roots at the edges of the monolith from being severed during soaking and the following washing procedure (Salonen, 1949; Fergedal, 1967; Kirby and Rackham, 1971).

The needleboard with the soaked monolith is then raised so that the water level of the tank is always 2–3 cm below the surface of the soil monolith, and tilted slightly to allow soil and water to flow away. With low water pressure of a hand sprinkler starting from the bottom of the board and working upwards to the plant stems, the soil is washed away. The strength and volume of the water stream must be controlled by a hand-operated valve. Too much pressure is to be avoided. Instead of using a hand sprinkler, a rotary sprinkler with three or four arms (Schuurman and Goedewaagen, 1971) or an oscillating sprinkler (Fig. 10.10) can be used.

Washing out soil monoliths with high clay content is extremely difficult, and often only possible with a considerable loss of roots. To facilitate the washing procedure and to minimize the loss of roots Schuurman and Goedewaagen (1971) recommend drying the entire monolith gently at 100° C, soaking it in a solution of sodium pyrophosphate (see Sect. 11.3) and then washing it as described above. Often this process of drying and soaking must be repeated.

Another technique to make the washing of monoliths with high clay content easier is to freeze them. The monolith is saturated with water in a tank, frozen to about minus 25° C, thawed in water and then washed out. According to Fergedal (1967) monoliths composed of heavy clay may require eight to ten alternative freezings and washings. It seems that the danger of destroying cereal roots by freezing the monoliths is slight; the washing away of the soil can cause more damage. This statement should, however, not be generalized, as roots of other plant species may be more susceptible to frost damage. Schuurman and Goedewaagen (1971) combined the method of freezing with the process of soaking in a solution of sodium pyrophosphate.

Fig. 4.11. Root system of a young tobacco plant obtained by the needleboard method. (De Roo, 1961)

4.6.5 Photographing and Sectioning

After the root system has been washed out completely, it can be photographed without removing it from the board. The best way to obtain pictures from the root system in its natural position is to photograph the needleboard under a thin layer of water because the finer roots are thus floating freely, and do not adhere to the main roots. A quicker technique is to take the needleboard out of the water and to place it against a vertical wall (Fig. 4.11). To keep the fine roots as nearly as possible in their natural position the board is taken out of the tank after its water level has been lowered slowly by syphoning.

Another possibility is to photograph the root system when it has been dried (Schuurman and Goedewaagen, 1971). The water is again carefully siphoned out of the tank, and with a fan the root system is blown air dry. It is then transferred on to a black hardboard for photographing. If a polyethylene sheet has been placed on the needleboard before excavating the monolith, the transferring of the dried root system is very easy. Although with this drying the finer roots are not as clearly visible as when they are photographed under water, the pictures obtained are satisfactory enough for demonstration purposes.

For ecological studies the quantitative root distribution in the different soil layers is more commonly determined by separating the root system into several sections. This can be done when the root system is still lying on the needleboard. Depending on the size of the board, the distance of the needles on it, and the amount of roots in the different soil depths a sectioning into 5- up to 20- cm layers is common.

If a separation into segments is not the research aim, the complete root system can be stored in its entirety after drying by being glued on the board. Preserving is

Fig. 4.12. Fitting the frame over a prepared free-standing soil monolith. (Method of Nelson and Allmaras)

also possible by placing it on a wooden board and by covering it with polyvinylchloride plastic from a spray can (Moraghan, 1968). This technique has, however, not been tested widely.

4.6.6 Special Modifications

Instead of excavating the soil monolith directly at the natural site with a needleboard this method can be modified in the following way. A soil monolith is taken out of the trench as described by the box methods (Sect. 4.4), and transferred to a strong screen with cross pieces studded with nails to hold the roots in place during washing. Sprague (1933) and Blaser (1937) using small monoliths about 10 × 20 × 30 cm and Spirhanzl (1936) and Najmr (1957) using large monoliths up to 50 × 100 × 150 cm have worked with this technique. The comments of the latter research workers on their experiences with this technique for large monoliths are, however, not encouraging.

A modified needleboard method for taking large soil monoliths which requires much less manual effort in the field has been introduced in ecological root research by Nelson and Allmaras (1969).

Fig. 4.13. Pushing long steel rods into a soil monolith through the perforated encasement board. (Method of Nelson and Allmaras)

Free-standing monoliths 152 × 41 × 91 cm were prepared by means of a small trencher having a 200-cm chain boom length which makes trenches 10-cm wide. First two parallel trenches were made which for the root study of row crops were laid out perpendicular to the row with the plants in the middle of the monolith. Then planks were inserted in these trenches along the outside edge of the surface plane. This kept the soil from filling the dug trenches while the next two trenches were made parallel to the row direction. Some additional work with a spade must be done to cut the soil monolith to the exact desired size. The depth of the trenches is determined to the maximum rooting depth of the plants.

A root extraction frame made from plywood is then inserted into the trench enclosing the whole free-standing monolith (Fig. 4.12). After the two sides of the frame have been fastened by nuts and bolts it is lifted out of the trench with a hydraulic tractor. To prevent loss of soil from the bottom of the monolith a piece of wood is placed against the bottom.

Through the back side of the frame, which has pre-drilled holes in horizontal and vertical rows like a needleboard, long steel nails or rods (400 mm long, 5 mm thick) were pushed by hammer or compressed air gun (Fig. 4.13). In this way the one side of the excavation frame is being prepared to function as a needleboard.

Fig. 4.14. Preserved root system of a soybean plant after the stage of flowering obtained with the method of Nelson and Allmaras. For better contrast the brown roots are placed against a white background

The monolith is then placed in a water tank at an incline to soak. The one side of the frame without the nails or rods is removed, then the monolith is washed as described in Section 4.6.4 until the root system is coalesced on the back plate. Finally, the nails or rods are removed, and the root system can be sectioned or preserved in its entirety after being transferred on to a good contrasting background (Fig. 4.14).

This technique has been used in the last few years by several research workers for root studies on agricultural crops (Allmaras and Nelson, 1971; Chaudhary and Prihar, 1974; Follet et al., 1974; Allmaras et al., 1975; Böhm et al., 1977; Sivakumar et al., 1977). The size of the monoliths in these investigation varied according to the plant species and stage of development. Although the excavation of the monoliths has been facilitated considerably, the general objections to the common needleboard method are, however, also valid for this modified technique.

Schuurman and Goedewaagen (1955) used the needleboard method to preserve soil profiles and their root systems simultaneously. With a common needleboard they took monoliths from the soil a little thicker than the length of the needles, and from this part of the soil outside of the needles they made a soil film. Then the remaining monolith was washed out, the exposed root system was dried, and soil film and root system were glued side by side on a cardboard.

4.6.7 Evaluation of the Methods

The needleboard methods have been found attractive by numerous research workers because simultaneously qualitative and quantitative studies of the roots are possible, and also the soil profile can be investigated easily. Most of the studies have been made with agricultural crops. In addition to the publications mentioned must be added the effective studies with cereals made by Goedewaagen (1941, 1955), van Lieshout (1956), Kähäri and Elonen (1969), with tobacco by De Roo (1957, 1960, 1961, 1969), and with potatos by De Roo and Waggoner (1961). For other plant species this method has seldom been used. Metsävainio (1931) studied bog plants, Portas (1973) vegetable crops, Coker (1958a) small orchard trees and brushes, and Armson (1972) conifer seedlings. For larger trees with their unelastic horizontal growing roots, this method is unsuited.

Compared with other root-study methods the needleboard method has the advantage that the cleaning procedure of the washed roots is reduced considerably. Many of the dead and old broken roots are washed away and mainly the live roots remain on the board. It should always be kept in mind that needleboard monoliths seldom contain the entire root system. Roots running towards the front and back of the monolith are cut off.

A disadvantage of the needleboard method is still the high labour requirement (Finney and Knight, 1973; Welbank et al., 1974). Therefore when labour is the limiting factor only needleboards up to a depth of 50 cm should be used. The method does not work in stony soils, and it will be problematic in very sandy soils, especially when dry.

5. Auger Methods

5.1 Specific Features

Of all sampling techniques, the auger methods are the most suitable for taking volumetric soil-root samples. The methods involve taking samples by hand augers or by mechanical sampling machines, and separating the roots from the soil by washing. Except for special purposes no attempt is made to keep the roots in their natural positions. The size of the samples generally is smaller than those from the monolith methods.

5.2 Sampling Techniques

5.2.1 Techniques with Hand Augers

5.2.1.1 Sampling Procedure

The simplest way of taking soil cores is by means of a hand auger. From the many types of augers the model developed by Goedewaagen is one of the best. It is described in detail by Schuurman and Goedewaagen (1971). Currently, this Netherland root auger is manufacturated commercially (Equipment for Soil Research B.V. Eijkelkamp, Lathum, The Netherlands).

The auger consists of a cylindrical tube 15 cm high with an inside diameter of 7 cm. Above the tube a hollow shaft about 100 cm long is fixed so it can be used for sampling up to a depth of 100 cm. On the outside of the tube and also on the shaft there are marks at 10-cm intervals. At the end of the shaft the auger is provided with a T-handle which makes it possible to rotate the auger while driving it into the soil and to pull out it again. The cutting edge of the cylindrical tube is serrated (Fig. 5.1). The hollow shaft contains a rod with a disc at the bottom which works as a plunger inside the cylindrical tube to force the soil cores out of the auger.

For taking soil-root samples the auger is pressed into the soil while being twisted, until the first 10-cm mark is reached, rotated several times and then pulled out. The easiest way to force out the core from the tube is to turn the auger upside down with one foot. Sampling is continued by replacing the auger in the borehole again and the whole procedure is repeated down to the next mark. When drilling in clay soils, the work can be facilitated if the auger is dipped into a pail of water before every drilling operation (Schuurman and Goedewaagen, 1971). If continuing investigations are required on the drilling site the holes are filled with soil immediately after sampling has been finished.

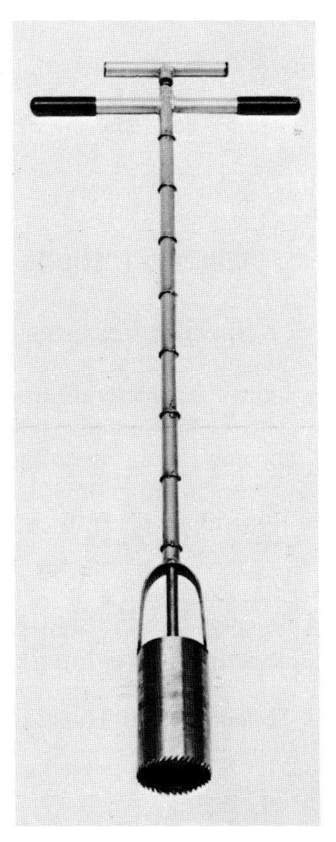

Fig. 5.1. Hand auger for taking soil-root samples to a depth of one meter. (With kind permission of Equipment for Soil Research B.V. Eijkelkamp, Lathum, Netherlands)

An extendable bi-partite model of the Netherland hand auger is also manufacturated by Eijkelkamp. This also works in hard soils, but a hammer apparatus is necessary for use with it. A modified type of auger with a serrated cutting edge was proposed by Borchert (1961).

Of the various other types of hand augers, that constructed by Albrecht has proved successful (Albrecht, 1951; Albrecht et al., 1953). The tube of the Albrecht auger consists of two halves held together by a metal ring, but which can be separated and removed from the shaft after removing the auger from the borehole. The side walls of the tube have longitudinal slots. As the auger does not have a serrated cutting edge, it is driven into the soil with a hammer. Several modified types of this Albrecht auger with varying tube diameters have been used by German scientists (Simon and Eich, 1956; Simon et al., 1957; Lippert, 1959; Skirde, 1971). As most of these studies were made on grassland, the sampling depth was seldom deeper than 50 cm.

The diameter of the core samples is a point of important consideration. Studies made by Schuurman and Goedewaagen (1971) showed that augers with cutting edges only 4 cm in diameter can cause too much frictional resistance between the core and the bore of the tube. It is true that friction also occurs with 7-cm cores but it can generally be neglected. Nevertheless, soil cores which are drilled with a 7-cm auger can be vary in length relative to the undisturbed soil (see Sect. 5.2.2). When sampling with an auger of a about 2 cm in diameter an inconveniently large number of replications must be taken (Angelo and Potter, 1939).

As recommended by Schuurman and Goedewaagen (1971) the most common diameter of soil cores for root studies with hand augers should be about 7 cm. Many research workers have used this core diameter (Williams and Baker, 1957; Crapo and Coleman, 1972; Fryrear and McCully, 1972; De Smet et al., 1972; Warsi and Wright, 1973; Richards and Cockroft, 1974; Köpke, 1979). Also when using mechanized sampling techniques a 10-cm core diameter is seldom exceeded.

5.2.1.2 Number of Replications

The question of how many soil cores must be taken at one site to obtain sufficiently reliable root data cannot be answered exactly, as recommendations derived from statistical calculations have seldom been reported in the literature. Most research workers have taken samples from three to six boreholes, similar to the common practice in field experiments for studying the aboveground parts of the plants.

According to the statistical calculations made by Opitz von Boberfeld (1972), who investigated grasses by taking samples with a diameter of 6.5 cm, about five replications are necessary to obtain significant results between varieties tested at 5% level by Duncan's multiple range test. He found differences, however, between grass species, and up to nine samples can be necessary.

Generally the cores should be taken randomly; but sampling on plots with cultivated crops growing in rows can be problematic. Here not only the number of samples, but also the location of the boreholes in relation to plants in the rows can be important. For root studies with cereals Schuurman and Goedewaagen (1971) recommended taking at least 24 borings per plot. Always half of the borings were taken in the rows, the other half between; but such a large number of replications needs a great deal of work which can seldom be done in practice.

Summarizing, it can therefore be concluded only as a general recommendation that with 7-cm augers samples should be taken from at least five boreholes. From the statistical standpoint no further discussion is needed that more than five replications are better, especially on plots with low rooting densities. The number of samples taken from each experimental unit is the primary factor governing the variability (Canode et al., 1977).

5.2.1.3 Special Technique for Studying Tree Roots

Hand augers can also be used for taking soil cores to determine the fine roots in tree stands, although because of the greater number of thick roots mechanized sampling techniques are generally to be preferred.

A good technique for studying the spatial root distribution of single trees has been proposed by Weller (1964, 1971) who investigated the distribution of root tips of orchard trees. At each chosen site he collected core samples from ten locations lying in a circle around the trunk, using a simple hand auger of about 4 cm diameter

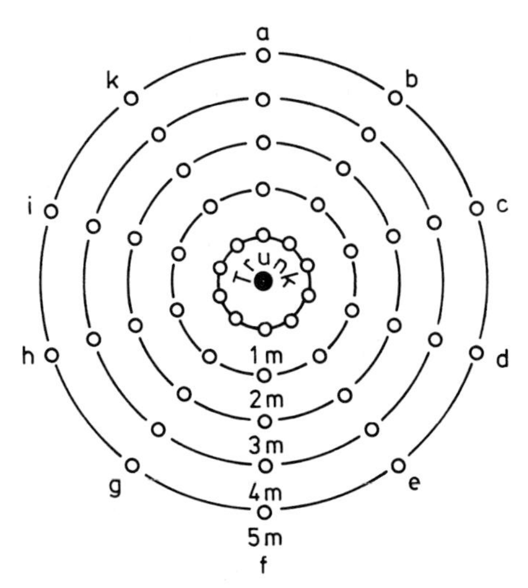

Fig. 5.2. Arrangement of sampling sites around a tree trunk. (After Weller, 1964, 1971)

and a special spiral drill to enlarge the sampling hole. He sampled in vertical steps of 10 cm to a depth of 2 m. This sampling was repeated in several concentric circles around the trunk (Fig. 5.2). The diameter of each circle was selected according to the nature of the tree under study.

In contrast to the excavation methods, the root system of the tree under study is not very much damaged by taking such small core samples with this technique. The technique is greatly appreciated by Russian scientists (Kolesnikov, 1971) and it should be used more often for studies on other tree species.

5.2.2 Mechanized Techniques

If core samples are to be taken below 1 m in depth, the use of hand augers becomes quite difficult and mechanized sampling techniques are to be preferred. However, even when the sampling depth is limited to 1 m, drilling by hand needs considerable strength to rotate the auger.

To avoid this heavy manual labour some types of augers are driven into the soil by means of a heavy drop-hammer and removed by a puller or a jack screw. One such type of soil sampler is the famous Veihmeyer tube, originally with a diameter of about 2 cm and a tube length in one section of about 180 cm, in some cases longer (Veihmeyer, 1929). It was used for root studies by Angelo and Potter (1939), Strydom (1964), Grimes et al. (1975), in a modified form with tubes of a diameter of 4.5 cm by Reynolds (1970), and with tubes 6.5 cm in diameter by Opitz von Boberfeld (1972), Opitz von Boberfeld and Boeker (1973), and Boeker (1974a, 1974b).

A pneumatic hammer (Kawatake et al., 1964; Bartos and Sims, 1974) or a portable petrol powered motor breaker adapted to drive coring tubes can also be used (Welbank and Williams, 1968; Finney and Knight, 1973; Welbank et al., 1974; Roberts, 1976). There are also powered machines working on the rotating hollow auger principle that are operated by one or two men (Buchele, 1961). However, these portable machines do not always perform satisfactorily for root studies.

Manual labour can be saved by using hydraulic, tractor-mounted core samplers. Boehle et al. (1963) used a steel tube 60 cm long and 15 cm in diameter as a sampler on a tractor drawbar which had hydraulic power for upward and downward movement. The heavy tractor gave the necessary weight on the drawbar to force the steel tube into the soil. Mitchell (1962), Bartos and Sims (1974), Mengel and Barber (1974a), Grimes et al. (1975), Kaigama et al. (1977) and Barber (1978) also worked with this technique for cutting soil-root cores, but using other dimensions of the sampling tubes. A widely used type of these core samplers is manufacturated by the Giddings Machine Co., Fort Collins, Col., USA.

Such mechanized techniques can cause more compaction of the cores in the tube than happens with the use of hand augers; but only few quantitative measurements of compaction in root studies are available. Welbank and Williams (1968) reported that by using a petrol-powered motor breaker the average compaction was not more than 2% to 3% during repeated sampling in spring and summer on a barley field. When taking samples from dry soils they observed no compaction. In moist condition the compaction of the core in the tube appeared to vary from nil in the upper soil horizon of 0–15 cm to about 5% in horizons below 30 cm. Welbank et al. (1974) found compactions of 25% and more when sampling a wet poorly structured soil.

5.2.3 Core-Sampling Machines

For agricultural engineering purposes including drainage studies and soil-survey mapping special machines have been constructed for taking long undisturbed soil columns (e.g., Bausch et al., 1977). Of the various types available the Kelley core-sampling machine which is also known as the Utah soil-sampler (Kelley et al., 1947) has proved the most successful for ecological root studies.

In brief outline, the principle of this sampling machine is as follows: with a set of three tubes an undisturbed soil column with a diameter of 10 cm is taken out from the soil plot to a depth of more than 3 m. The tube assembly consists of three tubes: one drive tube (outside tube) and two soil-carrying tubes (middle and inside tube). The inside tube is split longitudinally so that one half can be removed after sampling and the soil column can be obtained without damage. The outside tube which has a removable cutting edge, is the only rotating tube and is driven mechanically into the soil. The middle and the inside tubes move only in vertical direction. The inside tube has a "crowfoot head" which prevents the soil core from falling out of the tube when removed from the soil. A strong "driving chain" connects the set of tubes with a motor.

For extracting soil columns the sampling machine is brought in a vertical position exactly over the sampling position and the tubes are driven into the soil to the required depth (Fig. 5.3). After removing the tubes from the borehole, the middle and inside tubes are pulled out from the outside one, and then the inside tube is pulled out from the middle tube. Finally one half of the inside tube is removed and the undisturbed soil column can be divided into subsamples (Fig. 5.4).

Instead of sectioning immediately after sampling, the roots from the undisturbed soil column can also be washed out as a whole. In this case the core is

Fig. 5.3. Kelley core-sampling machine

wrapped in heavy aluminium foil and cradled in a semicircular through of sheet metal. At the washing place it is inverted on to a semicircular hardware cloth screen, soaked, and the roots are then washed free by a gentle flow of water (Kmoch, 1960a). After having been washed and cleaned, the root system can be transferred to a black background for photographing, and can then be sectioned according to the layers required.

Although the highly mechanized, man-power-saving procedure with this sampling machine seems attractive, the time needed for taking a column from one borehole is 10–30 min depending on the soil conditions. If the soil is too dry it can become impossible to obtain undisturbed soil columns longer than 1 m because the coring tubes can become blocked with compacted soil. Also compaction problems can occur similar to those with other core-sampling techniques (Sect. 5.2.2). Finally the drilling of one borehole disturbs the surrounding plants in an area of 8–12 m^2. Therefore in many cases, especially on small experimental plots, it is advisable to use hand augers instead of working with this type of sampling machine.

Root studies with the Kelley core sampler have been made in the USA by Ruby and Young (1953), Fehrenbacher and Alexander (1955), Kmoch et al. (1957), Bloodworth et al. (1958), Chandler (1958), Holt and Fisher (1960), Böhm et al.

Fig. 5.4. Sectioning the soil column obtained by the Kelley core-sampling machine into subsamples by means of a knife

(1977), and in Germany by Kmoch (1960a, 1960b, 1961, 1962, 1964a, 1964b), Kmoch and Hanus (1967, 1968), and by Hanus (1967, 1970). In most cases roots of agricultural crops have been studied.

5.3 The Core-Break Method

The time and labour needed for taking core samples in the field are relatively small compared with the additional work of washing and cleaning the roots and measuring the necessary root parameters. The washing and cleaning procedure can be avoided by a method where the soil cores taken out from the borehole are broken horizontally and the roots exposed at both sides of the breakage faces are counted.

This method of study was originally proposed by Hellriegel (1883), later used by von Seelhorst (1902) in Germany, and in a modified form by Fitzpatrick and Rose (1936) in the USA. In large-scale investigations Schuurman and Knot (1957), Vetter and Scharafat (1964), Vetter and Früchtenicht (1969), Ellis and Barnes (1977), and Köpke (1979) demonstrated the usefulness of this method.

Fig. 5.5. Breaking 10-cm long soil cores for counting the roots at the breakage faces

The cores are taken in the field by means of hand augers or mechanized samplers as described in Section 5.2, a proved length of soil cores being 10 cm. When they are broken in two approximately equal halves, the number of roots is counted at the breakage faces (Fig. 5.5).

At the two breakage faces it is not possible to distinguish whether visible roots are primary or secondary or belong to higher branching orders. Every exposed root—irrespective of its length—is counted as one root. The counting procedure is facilitated considerably if the two breakage faces are wetted with a fine spray of water which makes the roots more easily visible (Köpke, 1979).

Vetter and Scharafat (1964) and Vetter and Früchtenicht (1969) not only counted the number of roots but also measured their diameter. They then calculated root volume and root surface.

If only few roots are exposed at the breakage faces counting is very easy; but with increasing number, especially if highly branched roots occur, exact counting can become difficult and the true root number must be to a greater or lesser extent estimated. If too many roots are present in the core sample, counting can be avoided by comparing rooting densities at the breakage faces with specially designed standard figures (Schuurman and Knot, 1957; Schuurman and Goedewaagen, 1971). Such standard figures consist of a series of circles of the same diameter as the core with an increasing number of white dots on a dark background. Schuurman and Knot (1957) found that root coverage was correlated with root weight in the same cores.

All the known studies using the core-break method have been made with graminaceous species, and indeed, the method seems best suited for plant species with fibrous root systems. The counting procedure is simple and requires little time and labour. For correlations between root number and other root parameters see Section 12.2.

With the core-break method, which can also be regarded as a modified form of profile wall method, information about the approximate rooting densities in soil profiles is rapidly obtainable. Although the use of the method is limited when the soils are dry and the core samples crumble, it is not easy to explain why this field method has been used by so few research workers hitherto.

5.4 Advantages and Drawbacks

Core sampling in the field by hand augers or by mechanized equipment can be done reasonably quickly. When using hand augers the sampling can be done on very small experimental plots without significantly damaging the plants to be investigated. It is even possible to make a rough description of the soil profile by arranging the cores in order of increasing depth (Schuurman and Goedewaagen, 1971).

The relatively small diameter of the cores can be a disadvantage in sampling soil layers with low rooting densities. It can then happen—if the number of replications is not increased—that an unrepresentative picture of the root distribution in such soil layers is obtained (Kirby and Rackham, 1971). Core-sampling methods are not well adapted for morphological studies on roots.

It must always be kept in mind that with most of the sampling techniques the volume of undisturbed soil represented by the cores is not very exactly determined. Compaction occurs, and so the volume of undisturbed soil in the core can change with soil depth. Soils that are too dry and hard make the use of core samplers difficult, but if the experimental conditions allow, the drilling site can be rewetted one day before sampling. According to Kmoch and Hanus (1967) this has no significant influence on the root data obtained. In stony soils the core samplers do not work satisfactorily. Problems can also occur in tree stands because of difficulties in cutting thick roots.

In spite of these drawbacks the core-sampling methods are still among the most important in ecological root research. The limitation that it becomes increasingly difficult to take core samples from soil depths below 1 m is minimized considerably by using mechanized techniques. Probably also other mechanized soil-sampling techniques used for civil engineering purposes which are not mentioned here could be adapted for root studies.

6. Profile Wall Methods

6.1 General Survey

If plant roots are excavated from the soil it is in most cases sufficient to expose only one part of the whole root system. Such partial excavations as the sector methods (Sect. 3.3.4) can be modified by removing only few centimeters of soil from a profile wall and be used to record the exposed roots.

This was first done by Weaver (1919, 1920, 1926) who removed about 10 cm soil from smoothed profile walls with a scraper, and then made drawings from the exposed roots. Later van Breda (1937) and Blydenstein (1966) used air pressure, and Tharp and Muller (1940), Coetzee et al. (1946), Muller (1946), Davis et al. (1967), and Pinthus (1967) worked with a fine jet of water to expose the roots at the profile walls. But by removal of some centimeters of soil from the profile wall, the entanglement of the fine roots can be so confusing that only drawings, photographs, rough rooting density estimations, or verbal descriptions can be made. The information obtained with this bisect or transect method can be primarily qualitative.

To obtain quantitative root data, it is necessary to remove only a small layer of soil about 1 cm thick or less from the profile wall. The idea of scraping off only such thin layers of soil from smoothed profile walls and then counting the exposed roots was first proposed by Thiel (1892) in Germany. He tried to encourage farmers to look also at the belowground parts of the field crops in order to deduce practical consequences about manuring or soil management from the root distribution. But this appeal for doing root studies at profile walls found no echo. About 30 years later, Aaltonen (1920) in Finland counted roots in this way in forest stands.

The real acceptance of the traditional profile wall method came when Oskamp and Batjer (1932) made their intensive root studies on orchard trees at the New York Agricultural Experiment Station, USA. They marked the natural position of the roots at the profile wall on a chart by dots varying in size with root diameter. Since then, many research workers have used this profile wall method, at first only for studying tree root systems, later also for plants with fibrous root systems, such as grasses or annual agricultural crops. The method has been known mostly by the term trench profile method.

6.2 The Traditional Trench Profile Method

6.2.1 Digging the Trench

The position of the trench depends on the plant species to be studied. If plants in uneven stands, e.g., grassland, are to be investigated, the position of the trench is of

Fig. 6.1. Digging a trench with a digging machine for root studies with the trench profile method

no importance. On plots with plants growing in rows, the trench is generally dug transversely to the rows because in this case the root distribution between the rows and therefore the variation in the whole soil profile can be inspected much better. For studying tree roots, trenches are made 1–5 m or more from the trunk. For intensive studies trenches on four sides of a tree with the stem at the geometrical centre of the square are also possible (Garin, 1942).

The trenches are dug by hand or by a trench-digging machine (Fig. 6.1). The use of a modern mechanical excavation machine is a large help and saves many hours of work, but as also mentioned for the other root study methods, the machine will destroy much surface area at the experimental plots. Before using a digging machine it is necessary to mark the position of the profile wall by a spade-deep mini-trench. This will minimize possible destruction of the upper edge profile wall during the digging operation. In every case the profile wall should be positioned and prepared so that a further layer of soil 3–5 cm thick can be removed in the following process of smoothing to a perfectly vertical wall.

The length, width, and depth of the trench will mainly be influenced by the developmental stage of the plants to be studied and the aim of the research. Generally the working face at the profile wall should be 1 m in width, so the trench

Fig. 6.2. Smoothing a profile wall with a special blade

itself has to be about 1.20–1.50 m long. With trees or crops growing in widely spaced rows, a profile wall length of about 3 m is profitable (Oskamp, 1932, 1934, 1935a, 1935b, 1936; Oskamp and Batjer, 1932, 1933; Batjer and Oskamp, 1935; Havis, 1938; Levin et al., 1973). The depth should be at least 1 m, even if only the upper soil layers contain roots.

6.2.2 Preparing the Profile Wall

After the trench has been dug the final working face of the profile is smoothed. For this purpose a special blade (termed profile knife) about 110 cm long with two grips for easy handling has proved satisfactory (Fig. 6.2). It can have two sharpened cutting edges, the serrated one being better in hard soils than the untoothed one.

With such a profile knife, a line about 1–2 cm deep is first marked at the soil surface, marking the final wall of the soil profile to be investigated. The distance of this line should lie as near as possible behind the existing rough profile wall, so that not too much soil must be removed.

All the soil 1–2 cm in front of this marked line is then removed (Fig. 6.3). This procedure starts from the soil surface and is continued down to the bottom of the profile wall. The use of a plumb line can facilitate the work. Then the profile knife is used to remove the remaining 1–2 cm layer of soil. Also starting from the soil surface, the knife is moved from side to side like a saw. If the knife works too hard, the soil layer which is to be removed is too thick. The procedure must then be interrupted and the thickness of the surmounting soil layer must be reduced with a spade.

In this way a nearly vertical profile wall is obtained. It is easier to obtain an exact vertical wall if two supporting vertical laths are placed right and left of the

Fig. 6.3. Rough preparation of a profile wall with a spade for obtaining a nearly vertical working face

profile working face. The profile knife can then slide up and down on these laths. After some experience, however, these blade guides become unnecessary.

Finally, a trowel with a serrated edge can be used to give the vertical working face of the profile wall the final smoothing. For the subsequent exposing of the roots, the small rills made by this trowel are not a drawback, but actually beneficial. The use of such a trowel is preferred to an untoothed sharp metal sheet which can cover roots that are not cut clearly with the profile knife. After this smoothing procedure all the roots which protrude from the working face of the profile wall are cut.

At the bottom of the trench directly under the end of the working face of the profile, a small ditch about 10–20 cm deep is made. This ditch serves as the collecting place for the soil removed from the profile wall during the additional exposing procedure.

The smoothing of the profile wall works without difficulties in homogeneous soils. In stony soils the use of such a profile knife is not possible. Then the smoothing must be done with spade, small blades and different kinds of knives. If large stones are encountered at the profile wall which cannot be removed without destroying the surrounding smoothed working face, they are allowed to remain in place.

Fig. 6.4. Using a hand sprayer to wash a thin layer of soil away from a profile wall

6.2.3 Exposing the Roots

Generally a soil layer 3–5 mm thick is to be removed (Reijmerink, 1964; Kolesnikov, 1971; Böhm, 1976). Removing more than 5-mm of soil should be avoided because the time of exposing the roots is increased, as also that of mapping and counting. Only when working with thicker roots from trees can the thickness of the soil layer to be removed be increased to about 10 mm (Hausdörfer, 1959).

On homogeneous soils it is not very difficult to remove a soil layer of about 5 mm thickness. According to Böhm (1976) the mean variation in a homogeneous loess soil was between 2.5–7.5 mm. How accurately this exposing procedure can be done on other soil types has not been reported in the literature.

The roots are exposed with mechanical tools, by water, or by air pressure. The oldest technique is the dry procedure of exposing with screw drivers, long needles or by small-toothed scrapers. The procedure of exposing starts from the soil surface. Whether the exposing procedure is continued in one step to maximum rooting depth or is interrupted by interim mapping and counting of the roots, depends on the size of the profile wall and the kind of the roots. Roots large in diameter do not dry out so quickly if they are exposed; but for small fibrous roots, the exposing, mapping, and counting should be alternated at every 50-cm soil depth. On sandy soils it may be better to use a paintbrush for exposing the roots, perhaps in combination with other scraping tools.

For wet root exposing, a hand sprayer containing 5–10 l water kept under 3 atm of pressure is used. Also starting at the soil surface with a fine nozzle (Fig. 6.4), a soil layer of about 5 mm is removed from the profile wall. If the soil is too dry, the profile wall is pre-soaked with a fine spray of water, but without removing any soil. Where hard soil layers exist in the profile, the spraying procedure is aided

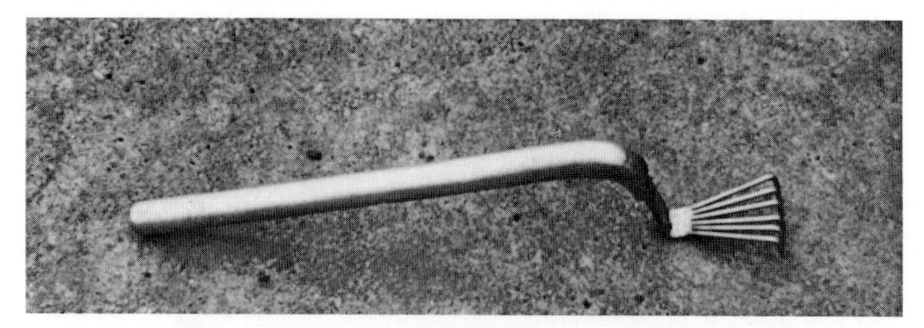

Fig. 6.5. Toothed metal scraper for loosening hard soil particles from a profile wall during spraying

by using a small-toothed scraper (Fig. 6.5). The scraper helps to loosen soil particles and to accelerate the spraying procedure.

To date, few researchers have reported using air pressure to expose roots at profile walls (see Sect. 3.3.2). This technique probably can be used successfully in sandy soils.

Summarizing the known experiences of research workers who investigated fibrous root systems with the trench profile method, it seems that the most efficient and quickest way of exposing is by a combination of spraying and scraping.

6.2.4 Mapping and Counting Procedure

6.2.4.1 Determination of Root Number

The mapping or counting of the roots should be done immediately after exposing. To facilitate this procedure a square grid net is placed against the profile wall and serves as a guide. The size of the grids depends on the size of the wall and the number of the exposed roots. For tree roots squares 10 × 10 or 20 × 20 cm in size are mostly used (Schuurman and Goedewaagen, 1971; Kolesnikov, 1971), for plants with fibrous root systems, such as grasses, a square of 5 × 5 cm has proved satisfactory.

The simplest way of covering the profile wall with a system of squares consists of driving nails in equal distances around the whole working face of the wall. Pieces of string, wire, or nylon thread are then stretched in such a way that a square grid system is obtained. This method has however, the drawback that the preparation of the grids must be renewed at every profile wall.

Much time can be saved if a frame with a complete system of square grids is fastened with nails to the profile wall (Fig. 6.6). The frame can be made from light metal or wood, the grid system consists of white nylon thread. A common size of such frames is 100 × 100 cm inner dimensions, but for investigating plants with many fibrous roots, a smaller frame of about 100 × 60 cm is more practicable.

The classical way of determining the exposed roots is by mapping them in their natural position on cross section paper (millimeter paper) with a scale of 1:10 or 1:5. At first, the soil depths are marked on the map, horizons are lettered and their boundaries are indicated by lines. Large stones or special soil features, such as

Fig. 6.6. Counting frame and other equipment used for the trench profile method (Böhm, 1976)

cracks and earthworm channels, can also be mapped (Scully, 1942; Melzer, 1962a).

The roots themselves are mapped by dots, each dot on the map corresponding to one root at the profile wall. With the aid of an ordinary ruler the mapping can be done with relative accuracy. If the exposed roots have different diameters they can be distinguished according to their thickness. The larger the diameter of the roots, the larger the size of the dot on the map (Fig. 6.7). Recommendable symbols for mapping roots with different diameters are presented in Figure 12.1.

The mapping becomes more difficult as the roots become smaller in diameter, especially if many branched roots have to be mapped. Most of the research workers do not mention this problem and it is seldom clear if they have marked only the main roots as one dot or if they have also marked all the branched roots as single dots. The dots on the maps can be counted later to obtain quantitative data.

Instead of dotting the exposed roots directly on a map and counting them later, the opposite technique is also possible. For this procedure only the number of the exposed roots visible in every grid of the frame is determined (Aaltonen, 1920; Aldrich et al., 1935). Root pictures can later be drawn from these data. Such pictures do not show the roots in their actual natural position because the total

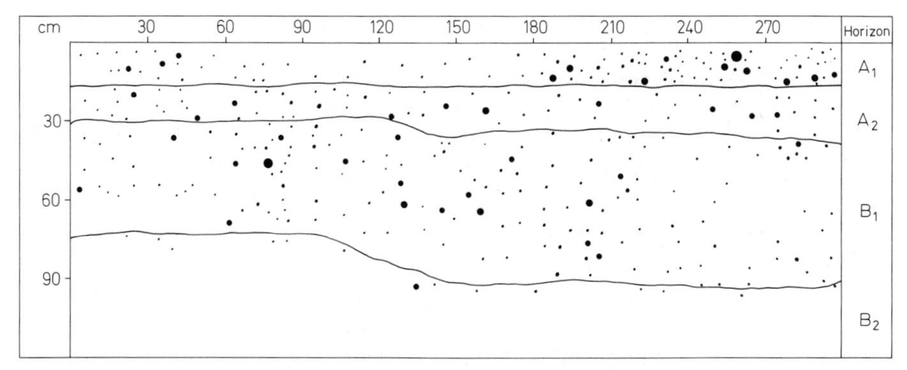

Fig. 6.7. Root distribution in an apple orchard. One of the first drawn root profiles obtained by the trench profile method. The larger the dots the larger the diameter of the exposed roots. (Redrawn after Oskamp and Batjer, 1932)

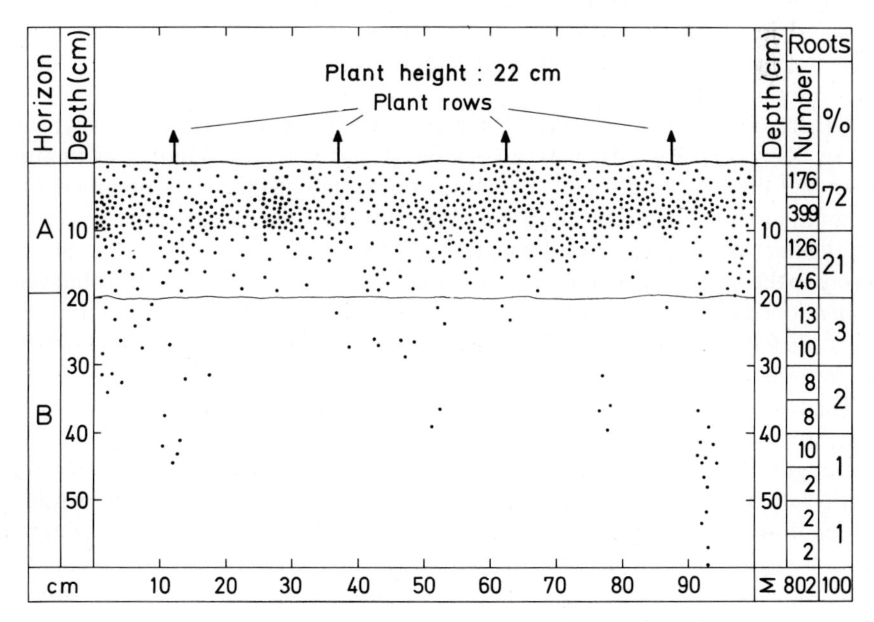

Fig. 6.8. Root profile of soybeans drawn according to the data recorded in the field. *Each dot* is 5 mm estimated root length at the profile wall. Root number is identical to 5-mm root-length units (Böhm, 1977)

number of roots counted in one grid later can be dotted only randomly in the appropriate grid area on the map (Bakermans and De Wit, 1970; Houben, 1972, see also Fig. 6.8). Different numbers of roots can also be combined into one category and maps with frequency classes can be made (Ehwald et al., 1955; Hausdörfer, 1959).

If the research aim is to obtain a picture of the actual natural root distribution in a soil profile, direct mapping in the field is preferred. If the root distribution is to be expressed only by number, the counting procedure should be used.

6.2.4.2 Determination of Root Length

Instead of counting root number, recently Böhm (1976) tried to estimate root length directly at the profile wall. The experiments were made with maize (*Zea mays* L.) on a loess soil. For the estimation of root length it was assumed that one root unit was 5 mm long. This means that every part of an exposed main or lateral root which was about 5 mm long was counted as one root. Roots which were 10, 20, or 30 mm long were counted as having 2, 4, or 6 such root length units.

Provided that the root distribution in the soil is nearly uniform, the approximate root length in a distinct soil volume can be calculated. If 5 mm of soil has been washed away from a 100 × 100 cm square profile wall, and if one estimated root unit is equivalent to 5 mm root length, a total count of 1000 root units in the whole profile corresponds to a root length of 1000 m/m³. Calculated on surface area it is similar to 10,000 km root length per ha.

To test the accuracy of this root-length estimation Böhm (1976) took soil monoliths from the same location, separated the root material by washing and measured the root length with a root-counting machine. The estimated root length at the profile wall was in all soil depths about one half of that measured with the root-counting machine.

Other comparative root studies made by Böhm et al. (1977) with soybeans and by Köpke (1979) with oats, also showed that with this kind of estimation about half the root length is to be found compared with the data obtained with monolith washing methods. The main reason for this underestimation seems to be that the fine roots which are still wet during the counting procedure often adhere closely to the larger roots so that they cannot be distinguished at the profile wall. As mentioned by Böhm (1976) and also found by Köpke (1979) some fine roots can be lost during the process of exposing, especially in hard and dense soils when a scraper must be used.

The few data available do not define a universal factor by which the root-length units estimated at the profile wall must be multiplied to obtain the approximate real root length. Probably, if not too many fine roots are involved, the estimated root length will be more in agreement with that obtained by a monolith washing method. Further investigations with other crops and trees are necessary to compare the root-length units obtained by this profile wall method with the data obtained by using the traditional sampling and washing methods. Root pictures can also be drawn from these estimated 5-mm root-length data (Fig. 6.8).

6.3 The Foil Method

A very good improvement of the traditional trench profile method was proposed by Reijmerink (1964, 1973), who studied the root distribution of asparagus. After the roots were exposed by spraying, he placed a transparent Plexiglas plate (5 mm thick) which was covered with a sheet of clear plastic foil (0.1 mm thick) in front of the profile wall. Then with a special pencil he traced the exposed roots as well as the limits of the soil and structural horizons on the plastic foil.

Fig. 6.9. Dotting root profiles in situ on transparent foil according to the method of Reijmerink. (Böhm and Köpke, 1977)

Later Böhm and Köpke (1977) slightly modified this method. They attached the transparent Plexiglas plate with supporting nails to the profile wall with a free space of about 2 cm between the wall and the plate. Instead of the clear plastic foil, they used a sheet of transparent acetate foil (0.05 mm thick) which was stretched between two rolls (Fig. 6.9). The dotting of the exposed roots was done with a water-proof felt pen. The Plexiglas plate was marked with a 5 × 5 cm grid net.

Dotting the roots on the foil takes no more time than is needed for counting the roots. As outlined by Reijmerink (1964) one person needs 1 h to trace 2000 roots on the foil.

Van Soesbergen (Soil Survey Institute, Wageningen, Netherlands) working on a silty loam attached a plastic foil directly to a profile wall as proposed by Breteler and van den Broek (1971) and mapped the roots by drawing as lines (Fig. 6.10). Attaching the foil directly to the profile wall, without having a rigid plastic or glass plate between foil and wall, should, however, be done only if the procedure for exposing the roots is made without using water. Otherwise the foil will become dirty and the exact mapping of the roots is impossible.

One drawback of the foil method is that the operator's direct contact with the exposed roots at the profile wall is lost. On looking through the foil—although the view is very clear—it can sometimes be difficult to decide if an exposed root is a root from the plants under study or if it is an old dead root from a crop of the previous year. The advantages of Reijmerink's foil method are, however, so striking that such small limitations can be tolerated. With no other root-study method can such natural pictures of the real root distribution in a soil profile be obtained.

The foil method is the most appropriate to investigate not only root distribution but also different diameter classes of the roots, or if the roots from different plant

Fig. 6.10. Drawing a root system of potato plants on plastic foil attached directly to a profile wall. (With kind permission of G.A. van Soesbergen, Soil Survey Institute, Wageningen, Netherlands)

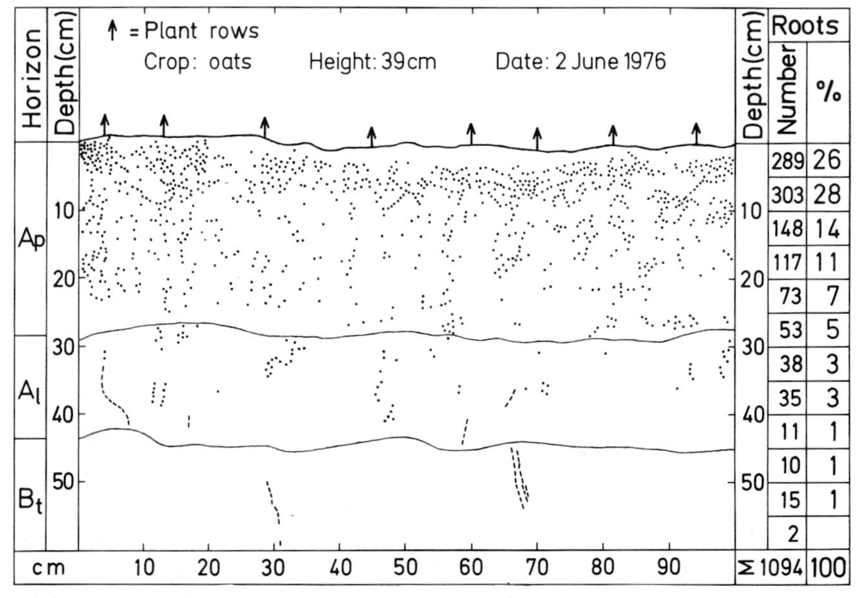

Fig. 6.11. Root profile of oats obtained by the foil method. *Each dot* is 5 mm estimated root length at the profile wall. Roots growing in earthworm channels are marked with *broken lines*. Root number is identical to the 5-mm root-length units. (Böhm and Köpke, 1977)

species have to be distinguished (Preston, 1942; Steubing, 1956, 1960). In these cases different symbols or coloured felt-tip pens can be used for drawing the roots on the foil. Böhm and Köpke (1977) differentiated between roots growing in the bulk soil and those growing in earthworm channels (Fig. 6.11). The relatively easy way to illustrate the real position of the roots in a soil profile with the foil is a significant advance in root ecology.

6.4 Technique in a Horizontale Plane

Instead of exposing the roots at a vertical profile wall, this can be done in a horizontal plane. Starting from a trench, adequate horizontal areas in different soil depths are prepared in a way similar to that described in Section 6.2.2. Generally the exposing of the roots should not be done by spraying with water because the horizontal plane will not allow the soil mud to flow away. Only if the horizontal plane is relatively small and a ditch has been made on the sides, is spraying advisable.

The exposed roots are counted on the cleaned horizontal plane also with the aid of a counting frame. Differentiation of roots growing in the bulk soil from those growing in earthworm channels seems to be more effective with this technique in the horizontal plane than with the normal vertical procedure.

The technique of mapping roots in a horizontal plane was used by Witte (1929) for studying grass roots and later by Russian scientists (Kulenkamp, 1969; Kulenkamp and Durmanov, 1974) for root studies on orchard trees. In a sense it is a modified form of the core-break method (see Sect. 5.3).

6.5 Evaluation and Applications

The determination of roots at profile walls, either if root numbers are counted or if root lengths are estimated, seems to be one of the best root-study methods from the ecological standpoint. The combined soil and root profiles obtained by this method are good basic material for interpreting root data, especially if results from different sites are to be compared.

To examine the root distribution of crops grown in rows, the trench profile method gives the best insights. Certainly, the root distribution between rows can also be determined by auger methods (Kaufmann et al., 1972) but it is more instructive to expose the roots at a profile wall.

The trench profile method is adapted for nearly all kind of soils, including stony soils. Problems can occur on soils high in clay content or in very loose soils. The use of the method is excluded if the roots themselves are to be collected for determination of dry weight.

Most workers who have determined root distributions in soil profiles have not done so with replications. Böhm (1976) made three replications with a second and third profile in the same trench always 20–30 cm beyond the previous profile. Köpke (1979), working with different cultivated crops, for the first time investigated up to twelve separate profile walls located randomly in each treatment.

Adequate recommendations of how many profile walls must be investigated to obtain statistically reliable results still cannot be given; it depends on the size of the profile wall and the rooting density of the actual site and its variation. Further studies are necessary. But as shown by the research workers cited below, many of the ecological research problems can be solved by working only on one profile wall.

Root studies with the trench profile method on orchard trees were made by Oskamp (1932, 1934, 1935a, 1935b, 1936), Oskamp and Batjer (1932, 1933), Sweet (1933), Batjer and Oskamp (1935), Havis (1938), Adriance and Hampton (1949), Butijn (1955, 1958), Olivin (1965), and Levin et al. (1973), and on forest trees by Turner (1936), Lutz et al. (1937), Garin (1942), Scully (1942), Ehwald et al. (1955), Hausdörfer (1959), and Melzer (1962a, 1963). Different grasses and cultivated crops were studied by Robertson (1955), Loeters and Bakermans (1964), Loeters et al. (1969), Bakermans and DeWit (1970), Houben (1972, 1974), Haans et al. (1973), Böhm (1976, 1977, 1978b), Böhm and Köpke (1977), Böhm et al. (1977), and Köpke (1979).

It should be mentioned that the trench profile method needs no expensive equipment. Therefore, especially at places where hand labour is still cheap, it should be preferred for solving many research questions.

7. Glass Wall Methods

7.1 Introduction

In glass wall methods root growth is observed or recorded through glass windows placed against the soil profile. Sachs (1873) working in boxes with refilled soil was one of the first who used this technique (see Sect. 10.3.3). Root studies in undisturbed soil profiles were made for the first time by McDougall (1916).

This type of in situ method with continuous observation and recording of roots behind glass windows has led to the construction of simple root cellars and modern underground root laboratories. The results obtained with these facilities have considerably accelerated progress in many fields of root ecology.

7.2 Glass-Faced Profile Walls

Generally root observation behind glass windows is done at a vertical profile wall. Therefore a trench has to be dug and the profile wall has to be smoothed as described in Section 6.2. The simplest way is then to place a glass plate directly at such a smoothed profile wall and to keep it in position by means of long pins although this procedure mostly leads to variable soil contact (McDougall, 1916; Woods, 1959; Howland and Griffith, 1961; see Fig. 7.1).

The size of the glass windows depends on the research aim, but they should not exceed 50 × 50 cm at a thickness of 5–10 mm. If the observation area must be larger, several windows are placed at the wall and they are connected by a rubber solution or an artificial resin and fitted into a frame. To minimize the danger of the windows' breaking, glass plates reinforced by an internal wire grid can be used (Rogers, 1939b; Fordham, 1972). Instead of glass also Plexiglas can be used (Fors, 1973; Fernandez and Caldwell, 1975, 1977).

Garwood (1967) installed glass plates in steel frames at the profile wall in such a way that the panels slid down within the frame so that their lower edges cut away a small layer of the exposed soil from the wall. This procedure ensures a better contact between the glass and the soil, although this can only be done in sandy soils. In clay soils, especially if they are wet, the smearing effects on the interior side of the panel will interfere with clear observation of the roots.

Good contact between the observation window and the soil is important. If free space exists between the window and the soil, the environmental conditions for the roots can be modified too much. Other drawbacks are that water can condense or soil can slip down into the free spaces, thus limiting the possibilities for observing the roots.

Fig. 7.1. Observation of root development of young forest trees grown behind glass-faced profile walls. (Howland and Griffith, 1961)

To overcome these difficulties, Rogers (1933, 1939b) carefully removed a thin layer of soil of some centimeters by horizons from the smoothed profile wall. This soil was spread out to become nearly air dry and sifted through a sieve to remove stones. Then the glass plate, faced in a strong steel framework, was placed a few centimeters away from the profile wall and the dried soil was packed between plate and profile wall. This repacking of the dried soil was done with a wooden stick layer by layer according to the natural soil horizons so that it occupied the same volume as before. If this is properly done, a good contact between the glass and the soil will be obtained. The natural soil moisture will gradually spread to the glass and the repacked soil will after some time have the same water content as the undisturbed soil profile.

This technique of repacking a thin layer of soil is now the most common procedure for installing root observation windows in the field (Pearson and Lund, 1968; Kolesnikov, 1971; Fordham, 1972; Durrant et al., 1973). It is also used for placing windows in the root laboratories (see Sect. 7.3.2).

Generally today, the observation windows are placed nearly vertical instead of on a slope against a soil wall. This is done to minimize not only the breakage of the glass plates, but also the difference in rooting density observed behind the glass and that present in the bulk soil. If only the development of single roots, their ramification etc. is to be determined, or if the root systems of plants in the seedling stage are to be studied, sloping windows are preferred.

After installation, the observation windows should be protected by a wooden or plastic plate to prevent light from reaching the roots (see Sect. 7.3.4). To maintain the temperature of the soil directly behind the glass window close to the natural soil

Fig. 7.2. General view of an observation trench beside an apple tree at the East Malling Research Station, England. A thermograph and a soil moisture meter are shown on the left. (Rogers, 1939b; with kind permission of the East Malling Research Station, England)

temperature, Roberts (1976) covered the observation windows with 5 cm thick sheets of expanded polystyrene. This precaution does not, however, always seem necessary because the temperature fluctuations directly behind the windows generally seem to be low (see Sect. 7.5). Direct sunshine on the windows should be avoided. In Northern latitudes the windows should face north, though it should be considered that north-facing windows may be difficult to observe with a binocular microscope in the absence of a good light source.

If root studies are to be done over a long period, the observation trench can be fitted with wooden or plastic plates to minimize danger of encaving (Ovington and Murray, 1968). The trench is covered with a roof made of asbestos or wood (van der Post and Groenewegen, 1960; Speidel and Weiss, 1974). Such a roof should slope gently away from the observation window to direct excess rain away. Another kind of roof construction used by Rogers (1934, 1939b), and Rogers and Booth (1960) is seen in Figure 7.2. During cold winters the roof should be covered with straw as an additional protection against frost (Göttsche, 1972) or the whole trench can be filled with straw (Kinman, 1932).

The techniques of observing and recording the roots at the glass-faced profile walls are the same as outlined in Section 7.3.3.

Continuous in situ observations at vertical profile walls without glass plates were made by Zillmann (1956). He dug an observation trench, and covered the observation area of the smoothed profile wall with wooden plates which he removed for observation and recording the roots. Drawbacks to this technique are that the roots visible at the profile wall can be damaged by the repeated removal of the wooden cover plates and that the contact between these plates and the soil is not always close enough for many of the visible roots to grow in a free space. Nevertheless, the data presented by Zillmann with several field crops show that good information about the seasonal root development is obtainable.

7.3 Root Laboratories

7.3.1 General Survey

The most modern development of the principle of observing and recording roots at glass-faced profile walls are the root laboratories. They are underground walkways having transparent windows on either side. The aerial parts of the plants under study are exposed to field environmental conditions.

The earliest root laboratories were constructed in Germany between 1900 and 1905 by Noll in Bonn, and by Kroemer in Geisenheim (Kroemer, 1905, 1918a, 1918b), and 1915 by Modestov in Russia (Kolesnikov, 1971). Another was constructed in Wageningen in the Netherlands (Blaauw, 1923). These early constructions consisted of an underground walkway with four to six observation chambers mostly containing artificial soil. These constructions at that time were termed root chambers, root cellars, or root houses. Similar facilities were later constructed for example, by Jacobs (1931), Könekamp (1934), Bosse (1960), Meusel (1961), and Fernandez and Caldwell (1975, 1977).

The two prototypes of the modern root laboratories were constructed in 1961 and 1966 at the East Malling Research Station in England (Rogers, 1969). These two underground buildings with 48 observation windows have influenced modern root research considerably, and have lead to the construction of many similar root laboratories in various countries. Depending on the research aims, the soil behind the observation windows in these new laboratories consists of undisturbed natural soil profile or chambers with definite artificial soil.

In the USA and recently more worldwide, these underground laboratories are mostly termed rhizotrons. At first the term was used by Lyford and Wilson (1966) for a special bulldozed root observation trench covered with a tent. The first modern root laboratory officially termed rhizotron was that constructed in 1967 at theUniversity at Guelph in Canada. Also the first root laboratory in the USA built at Auburn, Alabama was called a rhizotron from the beginning (Taylor, 1969).

7.3.2 Features of Their Design

The best examples of modern root laboratories with natural, undisturbed soil profiles behind the observation windows are still the two constructed in East Malling, England in 1961 and 1966. Features of their design are described in detail by Rogers (1969) and Rogers and Head (1963a, 1963b, 1968).

Fig. 7.3. Exterior view of one root laboratory at the East Malling Research Station, England

The two root laboratories are basically similar in their construction. At first a long trench was dug by a tractor with a mechanical scoop, leaving the soil profile on either side undisturbed. Then a framework of interlocking, pre-cast, reinforced concrete pillars and lintels was erected with a roof cast in situ on curved expanded metal reinforcement (Fig. 7.3). The length of the whole construction is about 30 m, the width 2.1 m, and the internal height from floor to ceiling also 2.1 m.

Between the pillars there are 48 observation windows (24 on each side), 122 cm deep and 101 cm wide, supported on low concrete walls cast in situ and held at the top by adjustable steel clamps (Fig. 7.4). Before the window frames were installed, the soil profile along each side of the trench was described and the profile wall was cut back to about 2–3 cm behind the window position. The soil which was cut away was saved in boxes according to the layers of the soil profile, air-dried and then replaced in the same order behind the glass to obtain a smooth observation surface without air gaps. The windows were here nearly vertical, with only a 2.5 cm forward slope.

The edges of the roof are only 25 cm above ground level to reduce the effects of changed microclimate on the plants under study. Gutters formed at the sides of the roof collect rainwater and lead it to soakaways at the end of the roof. The building is fitted with electric light and has six small light-trapped ventilators. It is accessible by stairs from an annexed instrument room.

The panels consist of plate glass 6.4 mm thick, engraved with 1.3 cm squares for better locating and mapping of the roots. In the 1961 model the window frames are welded from suitable steel L and T sections, galvanized after manufacture, and the glass is bedded on rubber strips. In the 1966 model the glass is held in place with wooden laths and metal spring clips which allow the glass to move as the soil

Fig. 7.4. General view of the interior of one of the East Malling root observation laboratories. Opened window on *left* shows cherry root growth. Also visible are a binocular microscope (*front left*) and time-lapse cine cameras (*mid left and near right*). All windows not being observed are kept shuttered. (With kind permission of the East Malling Research Station, England)

expands and contracts and yet still retain good glass-soil contact. Some of the windows are divided into many small removable panels. The smaller panels are less likely to break. A fine film of terylene (Melinex) is placed behind these panels to prevent the soil from sticking to the glass when the panel is removed. The terylene film can be cut easily when soil-root samples are taken behind these windows. All the windows are normally covered with light-proof shutters.

To obtain continuous information on the soil environment behind the glass windows, thermocouples are fitted through the glass at various depths and distances. Also soil moisture tensiometers or tubes for measuring water content with a neutron probe can be installed. Other soil properties such as soil air composition can be measured if desirable.

Root laboratories with undisturbed soil profiles behind the observation windows have also been constructed in Kenya (Huxley and Turk, 1967), and in Oklahoma, USA (Shoop, 1978). The root laboratory in South Africa (Glover, 1967) has an undisturbed soil profile along only one side of the walkway. The other root laboratories constructed in the last few years, although based in their fundamental construction on the designs of the East Malling prototypes, have disturbed soil behind their observation windows. For this purpose they have single compartments made of reinforced concrete in which the soil is filled according to

Fig. 7.5. Exterior view of the root laboratory at Ames, Iowa, USA

individual purposes. Typical examples for this type are the constructions at Guelph, Canada (Hilton et al., 1969), and at Auburn, Alabama, USA (Taylor, 1969).

As the tall entrance constructions to the walkways usually used by the first root laboratories built can influence the microclimate for the plants under study, later entryways have been built flat. So for example, the new root laboratory built in 1975 at Ames, Iowa, USA, has been inserted into an open field. As an additional precaution, at the time when the plants in the compartments are under study, similar plants are also grown in the surroundings of the root laboratory (Fig. 7.5).

The increasing interest in more knowledge about the movement of water, nutrients, and pesticides in the soil has lead to the construction of combined rhizotron-lysimeter research facilities. One of the most modern types has been built at Muscle Shoals, Alabama, USA, at the National Fertilizer Development Center of the Tennessee Valley Authority in 1973 (Soileau et al., 1974). Filter plates and vacuum water sampling equipment at the bottom of the root observation chambers evaluate the mobility of ions in the soil (Fig. 7.6).

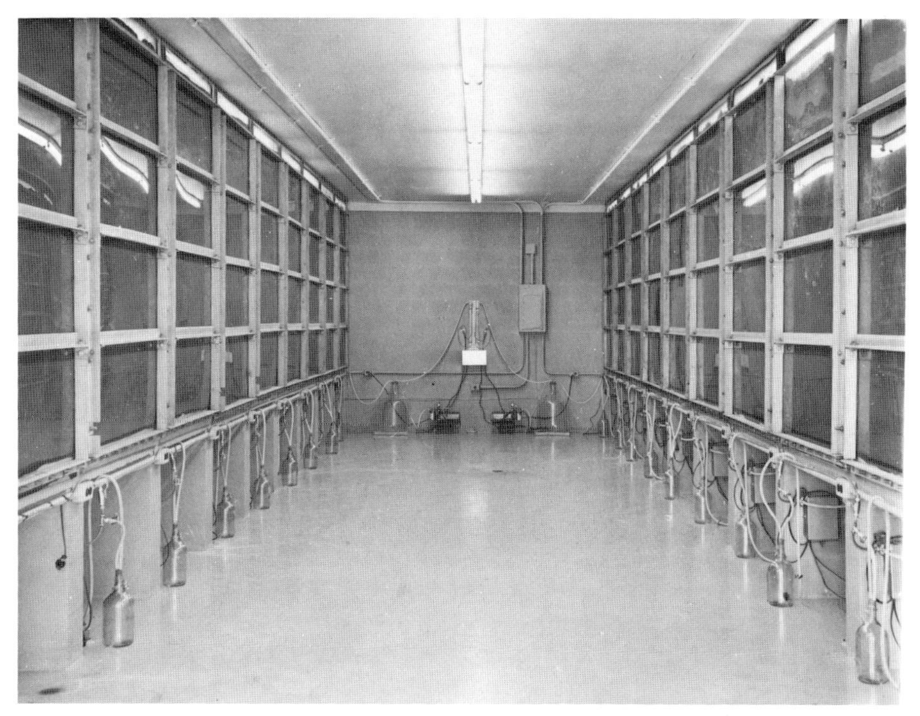

Fig. 7.6. Interior view of the rhizotron-lysimeter at Muscle Shoals, Alabama, USA. (Soileau et al., 1974) (With kind permission of the TVA, Muscle Shoals, Alabama, USA. Photo J.E. Soileau)

7.3.3 Methods of Recording

A more qualitative method of root observation involves making descriptions of root branching, root colour, direction of root growth, and similar features. Semiquantitative data can be obtained by visual estimations of rooting intensity based on a frequency scale.

For quantitative determinations the roots can be mapped on grid paper; but it is much easier and less time-consuming to place a transparent foil before the observation window and to trace the visible roots on the foil (Bosse, 1960; Meusel, 1961). For this purpose wax pencil or waterproof felt-tip pens are apt. If this root tracing is repeated at distinct intervals, different colours can be used (Kinman, 1932; Fordham, 1972). From this transparent foil root length can be measured by means of an opisometer, a rotating wheel used for measuring distances on maps (Ovington and Murray, 1968).

The quickest way to measure root growth behind observation windows is to count root intersections in situ on a grid system. This is now the most common technique for obtaining quantitative root data developed by Head (1966) and Newman (1966a). Details are outlined in Section 12.7.2. In practice the glass panels are engraved with horizontal and vertical lines, and every root visible behind the glass which is crossing a line on this grid system will be recorded. For plants with

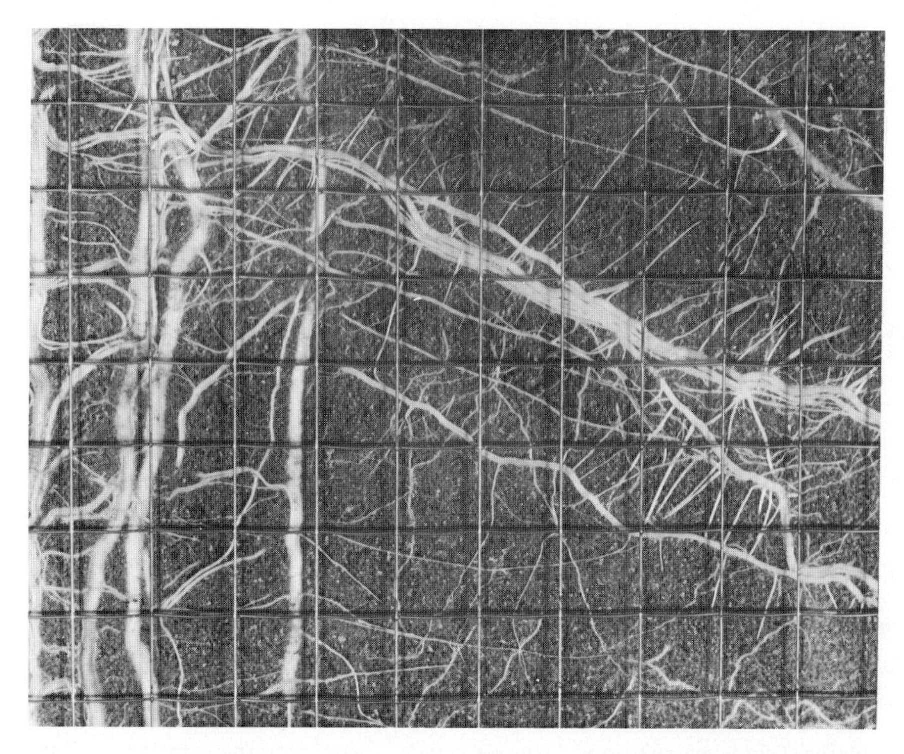

Fig. 7.7. Roots of maize (*Zea mays* L.) growing in a sandy soil behind an observation window in the rhizotron at Auburn, Alabama, USA. (Taylor, 1969)

predominantly vertical-growing root systems often it is sufficient to record only the roots which are crossing horizontal lines.

In most cases it is common to use a grid system of about 5 × 5 cm. However, the optimum square size will depend on rooting intensity. With very low intensities small squares are to be preferred; with high intensities even counting at 15 cm increments works satisfactorily (Taylor et al., 1970; Pearson, 1974). This intersection technique will give good information about the development of the root system and its distribution in the soil profile during a vegetation period of a plant.

The cm of root length visible per cm^2 of viewing surface is termed rooting intensity (Taylor et al., 1970; Klepper et al., 1973). Usually, rooting density, the cm of root length in a cm^3 of soil volume, cannot be calculated from the rooting intensity data, due to the root concentration effects at the glass surfaces. Such calculations can only be made as done in a loamy sand by Taylor et al. (1970), Taylor and Klepper (1973) and Klepper et al. (1973) if root concentrations at the glass surface are slight. These research workers assumed that they could record all the roots behind the observation windows in a soil layer of 2 mm thickness. Knowing the whole soil volume of the rhizotron compartment they calculated rooting density (cm root length per cm^3 soil volume).

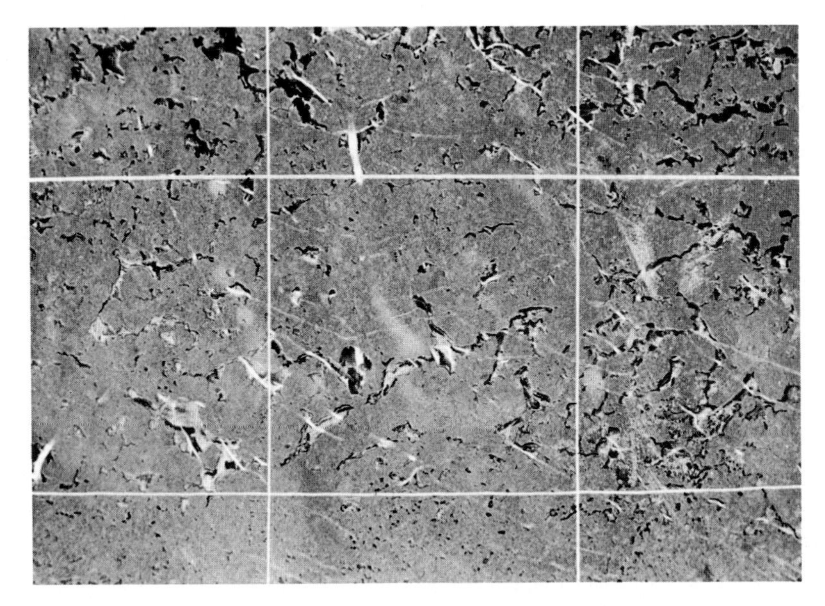

Fig. 7.8. Roots of soybeans [*Glycine max* (L.) Merr.] growing in a loess soil behind an observation window in the rhizotron at Ames, Iowa, USA

Before such calculations can be made, however, the amount of root concentration must be checked by taking also volumetric samples from the bulk soil behind the observation window, washing out the roots, and measuring their length. As found by Taylor and Böhm (1976), the root concentration can be considerably enlarged by using observation windows made from acrylic plastic and working in clay soils. If the clay soil becomes dry, cracks occur at the observation window interface, and many roots grow in these gaps.

In cases where a good contrast exists between white roots and dark soil at the observation window (Fig. 7.7), a photograph can be made, and the recording can be done by an image-analysing computer (see Sect. 12.7.3). Such good contrasts often do not exist however, and roots can be detected only with much effort (Fig. 7.8). If roots are brown, a bright–dark contrast, which is necessary for such a computer analyser, does not exist. So in most cases the visual in situ recording by an experienced operator will be the only practical technique.

The use of a straight binocular microscope enables detailed examinations of roots and soil fauna. The microscope can be placed on a wooden stand (Rogers, 1969) or mounted on a trolley which allows easy and very precise lateral and vertical movement while held constantly at right angles to the observation window (Bhar et al., 1969).

Valuable information on the development and decaying of roots can be obtained by using time-lapse cinematography (Head, 1965, 1968c). A normal cine camera can be modified for operation by an electric motor to take single exposures on a 16-mm film at time intervals regulated by an electric clock. The exposures are made by electronic flash light synchronized with the camera shutter.

7.3.4 Light Sensitivity of Roots

Light is necessary for the observation, counting, or measuring of roots behind glass walls. The question, therefore, is how the light during the recording time affects root growth. Rogers (1939c) found that continuous exposure to daylight severely checked root growth, hastened suberization, and mindered development of lateral roots of apple trees. Exposure to light for 20 min to 2 h per day caused some, but considerably less checking. At the weekly exposure of 30 min, the reduction in root length was statistically significant in the early summer when light intensity was high, but not significant in late summer and autumn. A typical negative heliotropic response was found in no case even on continuous exposure to daylight. Werenfels (1967) came to similar conclusions.

According to Pearson (1974), roots of maize, cotton, soybeans, and tomatoes show no difference in elongation rate on short illumination in front of the glass plates. In contrast, the root extension of peanuts was appreciable reduced by light. Hilton and Mason (1971) found that roots of poplar (*Populus alba* B.) grown in a rhizotron compartment suberized more quickly when exposed to natural and artificial light than when kept in the dark.

After prolonged illumination roots from some plant species become red or violet due to the production of anthocyanin (Siebert, 1920; Lyford and Wilson, 1966). Continuous exposure of light can also cause weak production of chlorophyll in roots of some plant species (Siebert, 1920).

Summarizing these results and observations from other research workers, the short time when the roots are exposed to light during recording in most cases has no strong influence on the results. As glass wall methods yield only relative data, so weak light effects during the short time of recording can be neglected in the solution of most of the ecological research questions. Keeping the windows dark at all the other times is necessary to prevent production of green algae, which can greatly reduce visibility through the windows.

7.3.5 Recent Research in Root Laboratories

At the root laboratories at East Malling, England, extended root studies on fruit trees and shrubs have been made. The main research was directed at seasonal changes in the quantity of the roots and their suberizing and decaying. Nowhere else, before or since, have such intensive studies on the root phenology of perennial woody plants been made (Head, 1965, 1966, 1967, 1968a, 1968b, 1968c, 1969a, 1969b, 1970, 1973; Rogers, 1971; Rogers and Head, 1963b, 1968, 1969; Atkinson, 1972, 1973, 1974, 1977).

Similar research on woody plants has been done at Guelph, Canada, including shoot–root relationship, diurnal variation in root elongation rates, and effects of soil temperature and soil moisture on root growth (Bhar et al., 1970; Hilton and Khatamian, 1973, 1974). Recently some agricultural crops have also been under study.

In the relatively small root laboratory at Eberswalde in the German Democratic Republic, built some years before the East Malling prototype was constructed, the shoot and root development of young forest trees was studied. Their growth rhythm data are valuable basic material for understanding the root phenology of woody forest plants (Hoffmann, 1966a, 1966b, 1967, 1968, 1972; Lyr and Hoffmann, 1965, 1967).

The growth of sugar-cane roots in relation to soils and climate was studied at the Mount Edgecomb laboratory in South Africa (Glover, 1967, 1968a, 1968b, 1970). The seasonal changes in extension growth of fine roots of *Arabica* coffee were investigated at the laboratory at Ruiru, Kenya, (Huxley and Turk, 1975). The periodicity of root growth of grapevines and its response to irrigation treatment was tested in the small laboratory at Griffith, N.S.W., Australia (Freeman and Smart, 1976).

The most intensive studies of annual agricultural crops have been done at the laboratory at Auburn, Alabama, USA. The relations between root distribution and water uptake by plant roots was the central research aim, but other studies such as observations of diurnal variations in root diameter and responses of different soil types to root penetration have also been done. Study plants were predominantly maize and cotton (Huck et al., 1970; Taylor and Lund, 1970; Taylor et al., 1970; Taylor and Klepper, 1971, 1973, 1974; Taylor, 1972; Klepper et al., 1973; Stansell et al., 1974; Huck, 1977).

At the new root laboratories constructed 1975 - 1977 at Ames, Iowa, USA, at Ibadan, Nigeria, and at Évora, Portugal, predominantly annual agricultural crops are under study.

7.4 Root Observations with Glass Tubes

Bates (1937) proposed another technique to observe plant roots in situ behind glass walls. He bored holes into the soil, forced tight-fitting glass tubes into these holes, and observed the roots visible behind the glass interface. He recommended this technique as a cheap alternative to the common glass wall observation trenches. The short communication of his idea unfortunately has been forgotten and has not been cited in any root publication since.

Later Waddington (1971) used a similar technique in a greenhouse experiment but working with a fibre optic for observing the roots. Inspired by this idea Böhm (1974b) tested this technique on a larger scale under field conditions.

With a blade auger, holes with a diameter of about 70 mm are bored vertically to a depth of about 110 cm. If in clay soils the blades of the auger cause smeared or compressed layers at the wall of the hole, a round steel brush is pushed up and down several times after boring. Then glass or Plexiglas tubes about 130 cm long and slightly smaller (60–65 mm) in diameter than the bore holes are inserted into the holes. The tubes are closed at the lower end by a rubber stopper. About 20 cm of the tubes remain above the soil surface. This part is wrapped outside with a foil or painted dark to prevent light from shining into the tubes.

Before inserting the tubes into the soil, a 5×5 cm grid system is marked by means of a black waterproof felt-tip pen or a diamond pencil on the external surface of the tubes.

To obtain good contact between tube and soil the free space between the outside wall of the tubes and the side of the holes is filled with dried, sieved soil from the

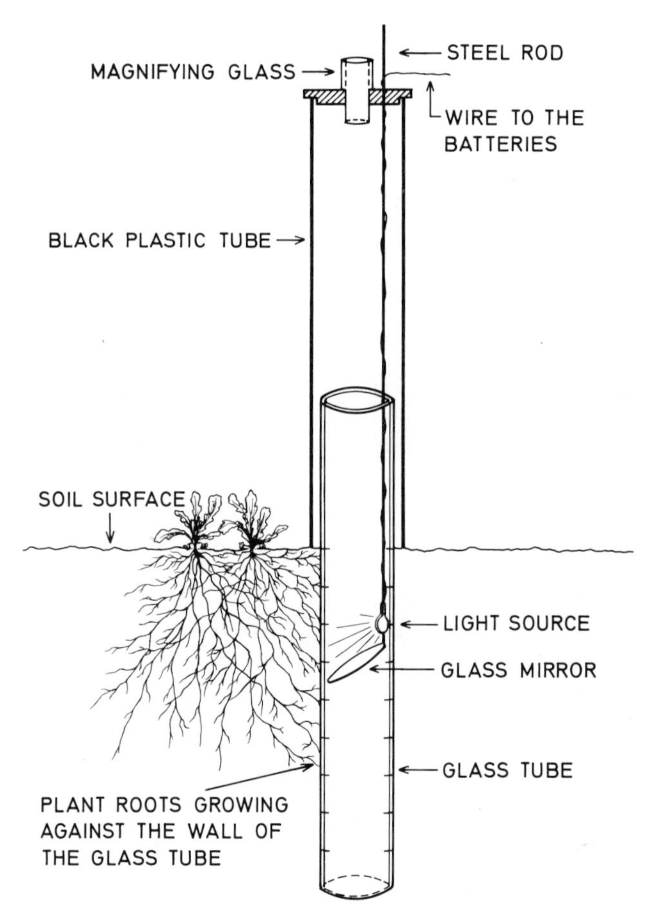

Fig. 7.9. Block diagram showing the principle of observing roots through glass tubes. (Böhm, 1974b)

same location. Gentle knocking on the part of the tubes above the soil surface avoids air gaps behind the observation squares. The top openings of the tubes are protected with aluminium foil or covered with a metal can.

After time has been allowed for natural consolidation of the refilled soil, seeds are placed around the tubes. For recording, the number of roots crossing the horizontal transects (in the case of many horizontally growing roots also the vertical transects) in every 5-cm depth are counted (see Sect. 7.3.3).

The roots visible behind the glass in the tubes are recorded with a round pocket-mirror and two 6-volt bulbs which are connected with two 6-volt batteries. The bulbs are fastened to a steel rod directly above the mirror (Fig. 7.9). Normally the roots can be counted with the naked eye to a depth of 1 m. The operator does not have to lie flat on the soil surface for counting, as a movable black tube about 30–50 cm high can be fitted over the observation tube.

The length and the diameter of the tubes can be varied for individual purposes. Böhm et al. (1977) used tubes 185 cm long and inserted these up to a depth of

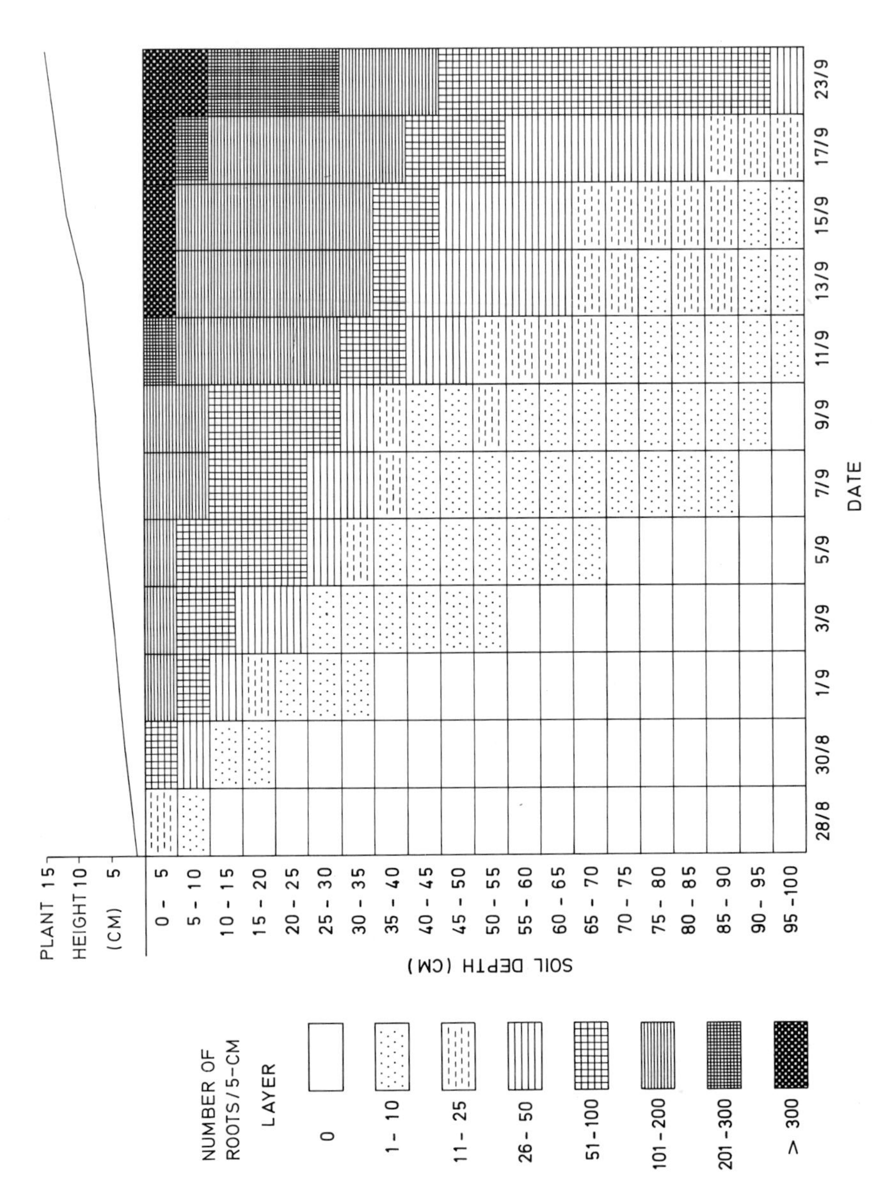

Fig. 7.10. Development of the root system of oil radish (*Raphanus sativus* L.) as measured by counting root numbers using five glass tubes. (Böhm, 1974b)

170 cm. If many roots are growing in deep layers it can be advisable to place a black plastic extension tube (about 130 cm long) over the tube to be recorded. Then the roots can be observed and counted with a fourfold magnifying glass (Fig. 7.9). If an additional nylon thread is tied to the steel rod, and a winch is put on the outside of the black plastic tube, the observation mirror can be moved up and down easily, and be stopped at distinct locations.

For observing roots in deep soil layers the light source should be increased. Böhm et al. (1977) fastened four 6-volt bulbs at the observation mirror for looking at roots deeper than 1 m. Although the light and mirror technique is simple, it works astonishingly well. Sanders and Brown (1978) modified this observation technique by using a highly refined fibre optic duodenoscope. They photographed the roots visible behind the glass tubes and made quantitative root length measurements from the photos.

Compared to the glass wall method at vertical trenches, the tube method is cheaper, causes less soil damage in the field, and the installation of a sufficient number of tubes also satisfies statistical needs. On the other hand most of the problems which are encountered with observing roots behind large glass windows in root laboratories also apply to the tube method. It is obvious that roots also concentrate behind these tubes. According to the detailed studies made by Köpke (1979) the root concentration in the 2-mm soil layer directly behind the tubes can be up to threefold that of the rooting density in the surrounding bulk soil. Presentations like those in Figure 7.10 show only relative root development in a soil profile. For solving many research problems, however, this method will be satisfactory, and in the last years many research workers have started to use this method. Böhm (1974b) termed the tubes which serve for root observations in natural soil profiles, minirhizotrons.

7.5 Evaluation of the Methods

The glass wall methods allow for a continuous study of the roots from one or more plants during their entire life span. Certainly, the roots are not growing in completely natural surroundings when they hit the glass panel and grow along it, but this does not seem to be as serious as might be thought. A glass panel can be considered to be like a large smooth flint stone or a grain of sand (Rogers, 1939b).

Of course there are also several problems associated with the glass wall methods. In the first months after installation of the observation windows the soil behind the panels can slake down so that air gaps can develop. This will be minimal if the windows are placed against natural soil profiles with only a small layer of soil repacked between the profile wall and the glass panel; but there can be a lot of soil settling if an entire rhizotron compartment has been refilled.

At the East Malling root laboratories, it was commonly found that root growth behind the observation windows was much greater in the first year after installation than in the following years (Rogers and Head, 1968; Rogers, 1969). In general, it is recommended that glass windows for root observations should be installed several months before the experiment starts.

If windows are properly installed they can work well for about 20 years or more; but as outlined by Kolesnikov (1971), it can happen that after a period of some years, the view through the glass becomes restricted by algae or hyphen. Soil animals or natural shrinking can also create gaps and channels in which water will condense, making observation difficult. Furthermore, as a result of continuous dying of the roots, an excessive humus layer can be established directly behind the glass windows which can create more favourable conditions for root growth. Therefore, in permanent root observation facilities, a 1–3 cm layer of the soil directly behind the observation windows should be removed and replaced with original soil from the same profile depths when it seems necessary.

It can be difficult to record roots behind the observation windows exactly if roots from the previous year have not sufficiently decayed; but this is at the same time an important advantage of the glass wall methods, that not only the growth of roots but also their decaying can be studied.

Temperature fluctuations directly behind the observation windows seem to be only slightly greater than in the bulk soil away from the windows. According to Rogers and Head (1968) at the East Malling root laboratory, the average weekly soil temperatures directly behind the glass and 1 m away from it in the bulk soil usually differed by less than 1° C.

One common drawback is that root development can seldom be recorded to the maximum rooting depth, as the observation surface of the windows is limited. Another limitation often is that the number of replications available for experimental treatments is too small, so that statistical calculations have not often been done by the research scientists working with glass wall methods. These difficulties may be overcome by repeating studies for several years.

Nevertheless, all these disadvantages are outweighed by the fact that continuous, exact determinations of the major changes in root behaviour can be made. The recording of roots behind glass walls is the most suitable study method to obtain knowledge of the root phenology of plants. Further root laboratories are planned and are under construction.

8. Indirect Methods

8.1 Introduction

The high input in time and labour required to obtain information about roots in the soil by direct observation or sampling methods has led to the development of indirect study methods. They are based on the principle of determining changes in water or nutrients in different soil layers between successive sampling occasions and from these changes inferring information on the root distribution in a soil profile.

Such indirect methods seem appropriate for ecological investigations, especially if the activity and not the absolute amount of roots in a soil profile is the primary research aim. Their success depends on several assumptions, which, if they cannot be met, may lead to erroneous results.

8.2 Determination of Soil Water Content

8.2.1 Gravimetric Method

This method is based on the assumption that correlations exist between soil-water depletion rates and the quantity of roots present in the soil. The effectiveness of the method was first demonstrated by Conrad and Veihmeyer (1929), Aldrich et al. (1935), and Veihmeyer and Hendrickson (1938).

For practical determinations soil cores are taken at different soil depths as described in Section 5.2. The diameter of the sampling tube can be smaller than 4 cm, and generally a hand auger will be adequate. The soil samples are dried to a constant weight at 105° C; soils with high content of organic matter should be dried at 50° C in a vacuum oven.

This appears an elegant method to measure directly how much water the plants have absorbed from the soil since the previous sampling occasion; but to interpret the results some assumptions have to be made. It is assumed that during the period of two soil water measurements there is nearly no transfer of water from one part of the soil profile to another, and there is no loss of soil water except by transpiration (Weller, 1964; Pearson, 1974). The water loss by evaporation does not seem to be a serious error if the soil surface is covered with a plant canopy. It is, however, important that the measurements should be made in relatively homogeneous soil profiles where discontinuity with gravel horizons or soil structure cannot affect water movement. Another assumption is that the period of water depletion commences with a uniform distribution of available water in the soil profile which approximates to field capacity.

It is impossible to obtain good results, especially in the upper soil horizons, if rain occurs between two sampling occasions. As shown by Böhm et al. (1977) the gravimetric soil-water depletion method was completely unsatisfactory for determining rooting density in an experiment with soybeans in Iowa (USA). During a very wet June the entire soil profile remained near field capacity, and the movement of water from one soil layer to another predominated over removal of water by roots within any soil layer. Thus, gravimetric samples taken at successive intervals did not show changes in water content although roots were active in these soil layers.

Nevertheless, other research workers under more suitable conditions found high correlations between water depletion and rooting density (Gliemeroth, 1952; Bennett and Doss, 1960; Doss et al., 1960; Linscott et al., 1962; Patel, 1964; Davis et al., 1965).

Summarizing the experiences with this water depletion method for measuring rooting density in the field it must be stated that direct study methods are preferred. Measuring water depletion should be used as a complementary method to obtain information on water uptake by roots in different soil layers (Hanus, 1962; Ehlers, 1975).

8.2.2 Neutron Method

The neutron scattering method is based on the principle that fast neutrons are slowed down and scattered much more by hydrogen atoms than by other atoms. As the concentration of hydrogen atoms in a soil profile is much higher in soil water than in the inorganic compounds and in the dead or living organic compounds, the method can be used for determining soil water content.

Access tubes made from aluminium or another metal are installed in the soil on the experimental site. A probe with a radiation source and a detector are inserted for each measurement into the tubes and the rate of slow neutrons is recorded. When the probe with the radiation source is out of the soil it is withdrawn into a shield, giving protection to the operator.

The number of neutrons attentuated in the soil is proportional to the volumetric water content. Since, however, the count rate recorded is not linearly related to the water content suitable calibration is necessary. High content of organic material, atoms of boron, iron or chlorine can falsify the results. There are several types of neutron instrument which have been mainly used for soil water studies without reference to the root distribution in the soil. Further details and a survey of the abundant literature are given by Slavík (1974).

The neutron method as an indirect method for estimating rooting density in soil profiles was used by Cahoon and Stolzy (1959), Haahr (1968a, 1971), Levin et al. (1973), Stone et al. (1976), and Castle and Krezdorn (1977). In spite of some good agreements between water extraction and root data obtained by direct study methods, the same assumptions and drawbacks outlined for the gravimetric method also apply to the neutron method. In addition, by the neutron method the limitations of the measurements in layers 0–15 cm close to the soil surface should be mentioned. Although appropriate correction factors can be applied, the information obtained should be interpreted with caution.

A new development is the use of neutron radiography for studying root growth in bulk soil. In laboratory experiments Willatt et al. (1978) found that neutron scattering by the roots allowed elongation rates of soybean radicles and roots of maize to be determined easily through either 2.5 or 5 cm thick soil samples. Although for this technique a nuclear reactor is used, it seems possible that a portable neutron generator for field studies will be available within some years.

8.3 Staining Techniques

The idea of injecting coloured liquids into a plant stem and observing the translocation of the colour through the parts of the plants is not new. As early as the fifteenth century liquid plant injections were made, and it is a common method in plant physiology today (Roach, 1939; Gurr, 1965).

Staining techniques are limited in applicability to tree root studies, e.g., for detecting natural root connections between roots of trees of the same species or between trees of different species (Graham and Bormann, 1966). The occurrence of natural root grafting in most cases has been detected with 0.1%–0.5% aquaeous solution of eosin red or eosin blue. A stem was selected as the donor and the aqueous dye solution was injected. Root grafting was presumed to exist between the donor and any stem which later became coloured. The dye absorption could be easily observed by removing the bark from the stem (Yli-Vakkuri, 1953; De Byle, 1964; Gifford, 1966).

Bormann and Graham (1959) and Graham (1960) developed a technique in which the donor tree was felled, a collar was constructed around the stump, and the dye solution filled into the collar. Removal of the crown of the donor tree ensured maximum potential movement of the dye into the root system.

Unfortunately, however, there are too many limitations to the staining techniques for detecting root connections, and the results must be supported by excavation and visual inspection. Problems are greater with increasing age of the trees. According to Graham (1960) in many cases the dye moved only short distances along roots of donor trees and then stopped inexplicably. So far, the staining techniques seem to be only of little value and are of very uncertain help in ecological root research.

8.4 Uptake of Non-Radioactive Tracers

Instead of injecting the plant stems with dye solution and then detecting the dye in the roots, the contrary technique is also possible. Tracer material which has been placed into the soil and taken up by the plant roots can be detected in the upper parts of the plants. The salts of lithium, boron, or strontium habe been used.

Sayre and Morris (1940) placed small quantities of lithium chloride in the soil and marked the locations between the rows of maize. Later they tested the leaves of nearby plants spectrographically for the presence of lithium. The data obtained were recorded only in a qualitative manner as positive or negative, indicating the presence or absence of the tracer. Fox and Lipps (1964) also tried using lithium for root studies of alfalfa, but they found this tracer toxic at rather low levels.

Hammar et al. (1953) used the uptake of boron as a criterion to estimate the relative spread of the roots from pecan trees. They applied borax to the soil around the trees by broadcasting and subsequently analysed the leaves for boron.

Fox and Lipps (1964) used stable strontium chloride, which was placed in the soil at various depths, for measuring root activity of alfalfa. The strontium taken up by the plants was later determined in the ashed plant material by a flame spectrophotometer. One advantage of strontium as tracer material is that in many soils it is not very easily leached from the point of application.

The results of the few experiments with non-radioactive tracers under field conditions do not, however, recommend these techniques today for routine studies in ecological root research. One point is that the uptake of one ion species will not necessarily reflect the activity of the root system for other ions (see Sect. 8.5.5).

8.5 Radioactive Tracer Methods

8.5.1 General Survey

Methods with radioactive isotopes have gained significance in ecological research during the last decades. The development and activity of plant root systems in a natural soil profile was first measured with a radioactive tracer by Lott et al. (1950) and by Hall et al. (1953). They placed ^{32}P in the soil at various positions and depths around grapes and maize plants, and the amount of radioactivity taken up by these plants was used as a measure of the intensity of root activity. Later Racz et al. (1964) reported on a technique where radioactive tracer was injected into the plant stem and the pattern of root activity was determined by taking soil-root samples and recording the radioactivity in them. These two approaches are now the common radioactive tracer techniques for ecological root studies under field conditions.

The most commonly used isotope for plant root studies is ^{32}P. It has a half-life of 14.3 days, is mobile enough in plants to become rather uniformly distributed throughout the root system in a short time, and is relatively inexpensive. For these reasons most of all the research workers cited in Sections 8.5.2 and 8.5.3 have worked with ^{32}P.

Besides ^{32}P ^{86}Rb also is of significance for root studies because it is a strong gamma emitter and decays fast enough (half-life 18.7 days) to minimize hazard when it is used under field conditions (Staebler and Rediske, 1958; Bormann and Graham, 1959; Trouse and Humbert, 1961; Saíz del Río et al., 1961; Russell and Ellis, 1968; Ellis and Barnes, 1973; Gerwitz and Page, 1973; Huxley et al., 1970, 1974; Lupton et al., 1974).

On a small scale ^{131}I has been used for tree root studies (Bormann and Graham, 1959; Price, 1965; Hough et al., 1965; Ferrill and Woods, 1966). Other root studies have been made with ^{42}K (Ozanne et al., 1965), and ^{45}Ca (Woods and Brock, 1964). But the use of these tracers has not proved very successful, partly due to high costs for the isotopes, or problems with the measurement technique. More promising seems the use of ^{137}Cs (Olson, 1968) or ^{134}Cs (Cox et al., 1978), especially for estimating turnover rates of roots. For techniques with ^{14}C see Section 8.5.4.

8.5.2 Soil Injection Technique

The principle of this technique consists of injecting a small amount of tracer solution (carrier free tracer or fertilizer labelled with isotopes) into the soil and measuring the radioactivity in the aboveground parts of the plants.

When using this technique to obtain information on the relative root activity in a soil profile, it is assumed that the roots have an equal chance of encountering the tracer at each spot injected, that the tracer is immobile in the soil but remains available for root absorption, and that the specific activity of the tracer is nearly uniformly injected (Pearson, 1974). ^{32}P is the most suitable tracer to meet these assumptions.

The injection spots are generally chosen in a concentric circle around the plant, or in a row at different distances from the plant stem and at different depths in the soil profile. In the simplest manner holes are drilled to the desired depth with a soil auger and labelled phosphorus as solid superphosphate is inserted to the bottom of each hole. The application is made through a tube so that its lower end is just above the bottom (Lipps et al., 1957). In most of the experiments tracer solutions have been used.

Precautions are necessary during the placement of the tracer solution into the holes. In every case drops of the solution must be prevented from falling on the plants (Fig. 8.1). To avoid the difficulties of contamination, leakage, and severe radiation exposure associated with the handling of tracer solutions, Jacobs et al. (1970) proposed placing the radioactive solution in gelatine capsules and freezing it quickly over dry ice. These frozen capsules are handled with forceps and dropped to the bottom of the holes through tubes. Due to the advantages of this technique, more and more research workers have used them in the last years.

After the tracer is placed in the bottom of the holes, these are refilled with soil from the same location. This refilling must be done carefully, with no air gaps, because otherwise roots can grow in these unnatural pathways without mechanical impedance. To overcome this difficulty Boogie et al. (1958), and Boogie and Knight (1960) placed the tracer horizontally from a trench running parallel to the treatment site. With a special tube similar to a large hypodermic syringe the tracer solution was inserted to the required depth and the trench was filled in. Freytag and Jäger (1972) proposed a technique of inclined injection of ^{32}P solution for tracing root growth, but the vertical placement by making holes to different depths with the auger is still the most common procedure.

The amount of radioactive tracer injected into every hole varies extremely. Generally it must be considered that with increasing levels of non-radioactive phosphorus in the soil profile the detectibility of the ^{32}P in the plants becomes lower. In cases of high levels of non-radioactive phosphorus in the soil, it is advisable to increase the quantity of ^{32}P or to use another isotope as tracer. Haahr (1975), who studied the root activity of barley plants, placed 0.4 ml ^{32}P solution in every hole with an activity of 100 μ Ci/ml in a frozen gelatine capsule.

Fig. 8.1. Injection of radioactive ^{32}P solution into the soil for determining root activity. (With kind permission of Letcombe Laboratory, Wantage, England)

Measuring the presence of ^{32}P in the aerial parts of the plants depends on the conditions of the experiment. If studies are made with isolated plants, a Geiger-Müller counter can be placed against the leaves and the reading can be recorded on a portable ratemetre (Boogie et al., 1958; Pettit and Jaynes, 1971). However, with this in situ measurement technique several problems are involved. Thus, for example, with time the radiation will become so weak that it cannot be detected above background radiation.

The better technique is therefore to sample periodically young leaves of the plants which are dried, ground, ashed at 500° C, and extracted with hydrochloric acid. The activity of this liquid is then determined with appropriate Geiger-Müller equipment, but also the dry ash can be measured directly. Huxley et al. (1974) compared liquid versus dry ash counting, and found that with standardization of the techniques, the counting of dry ash was quicker and more efficient.

Most of the root activity studies with the soil injection technique using ^{32}P have been made with cultivated plants. Predominantly annual agricultural and vegetable crops have been the subject of study (van Lieshout, 1960; McClure and Harvey, 1962; Hammes and Bartz, 1963; Nakayama and van Bavel, 1963; Haahr et al., 1966; Haahr, 1968b, 1971, 1975; Bassett et al., 1970; Agrawal et al., 1975; Paasikallio, 1976). Numerous studies have been made with grasses and other pasture crops (Burton et al., 1954; Lipps et al., 1957; Fox and Lipps, 1964; Lipps and Fox, 1964; Evans, 1967; Bray et al., 1969; Osman, 1972). Studies with tree

Fig. 8.2. Injecting radioactive tracer solution with a syringe into the stem of a cereal plant. (With kind permission of the Letcombe Laboratory, Wantage, England)

crops, especially those grown in the tropics, like coffee, cocoa, mango, palms, and citrus, have been started on a larger scale during the last decade (Ahenkorah, 1969; Forde, 1972; Huxley et al., 1974; Bojappa and Singh, 1974; International A.E.A., 1975).

8.5.3 Plant Injection Technique

In this technique the radioactive tracer is injected into the plant stem. After allowing time for the tracer to distribute throughout the whole plant, soil-root samples are taken, and the tracer is measured in them. The amount of radioactivity gives a measure of the amount of active roots in that part of the soil profile where the core samples have been taken.

With ^{32}P the technique was successfully demonstrated by Racz et al. (1964), and later used by Subbiah et al. (1968), Katyal and Subbiah (1971), Soong et al. (1971), Maurya et al. (1973), Mohr (1975), Vijayalakshmi and Dakshinamurti (1975, 1977), and Cholick et al. (1977). In most cases cereal crops have been investigated.

Usually, ^{32}P solution in dilute hydrochloric acid was injected between the first and second node of a plant (Fig. 8.2). Subbiah et al. (1968) used two calibrated microliter syringes, one for creating suction, and the other for injecting the ^{32}P-solution. The injection holes were sealed with collodium. Mohr (1975) worked with a wet cotton wick which he pushed through the plant stem by means of a needle. One end of the wick was dipped in a small tube containing the tracer solution, and in this manner the tracer was transported with the sap stream into all parts of the plant.

It takes between 5 h and 5 days for the ^{32}P tracer to equilibrate throughout the whole plant system. The speed of the translocation depends on the development stage of the plant, and is mainly influenced by the temperature. There are also large differences between plant species.

After the equilibration period, soil-root cores around the injected plant are sampled with a tube auger at various depths and lateral distances. The cores are dried, ground, and ashed at 500° C to ensure thorough mixing of the ^{32}P contained in the roots with the soil material. A 10-g portion of the ashed sample is compressed into a briquet and the activity is recorded using a Geiger-Müller counter. The activity at each location is expressed as percent of the total count from all locations. The results must be corrected for background radioactivity and decay of the tracer.

Russell and Ellis (1968) and Ellis and Barnes (1973) worked with this technique of plant injection but using ^{86}Rb. They found in graminaceous plants, 24 h after ^{86}Rb has been injected into the base of the stems, that its distribution is adequately uniform throughout the root system. During this period they found only a negligible outward diffusion of the tracer.

The main difference between using ^{32}P and ^{86}Rb is that the latter is a gamma emitter and therefore with suitable counting techniques there is no need to ash the soil-root samples. Furthermore the size of the samples counted can be up to 3.0 kg (Lay, 1973). Intensive use of the plant injection technique with ^{86}Rb was made by Lupton et al. (1974) for studying the root activity of different wheat cultivars. Gerwitz and Page (1973) used ^{86}Rb for root studies on vegetable crops.

Staebler and Rediske (1958) injected forest trees with ^{86}Rb and measured the radioactivity directly in situ at the soil surface with a portable scintillator. By drawing radiation isolines around trees based on counts per minute they obtained a rough picture of their rooting density.

Roots labelled with radioactive tracer by stem injection can also be detected in situ by driving an autoradiograph filmholder into the soil profile, or by placing it directly at a smoothed profile wall. By this technique the labelled ends of the cut roots appear after adequate time of exposure as dark spots on the film (Baldwin et al., 1971; Mohr, 1975).

This technique—like the common profile wall method—permits the study of the spatial root distribution in a soil profile, but there are still several disadvantages. Thus if high root concentrations occur in a soil layer there can be large black dots on the film, and single roots cannot always be detected separately (Fig. 8.3). On the other hand the activity of the fine roots is not always strong enough to show up on the film. However, further improvements are certainly possible. The injecting of various plants with different isotopes seems a suitable technique for studying interpenetration of root systems separately and in relation to each other (Baldwin and Tinker, 1972).

The plant injection technique does not show the ability of a root system to absorb nutrients from the soil profile at any given time. For example, when soil moisture is a limiting factor, the tracer injected into the plant stem can still be transported into live, but dormant roots (especially in tree-root studies) which are unable to absorb nutrients due to moisture stress (International A.E.A., 1975). Therefore if nutrient or water uptake in relation to root activity is to be studied the soil injection technique seems more appropriate.

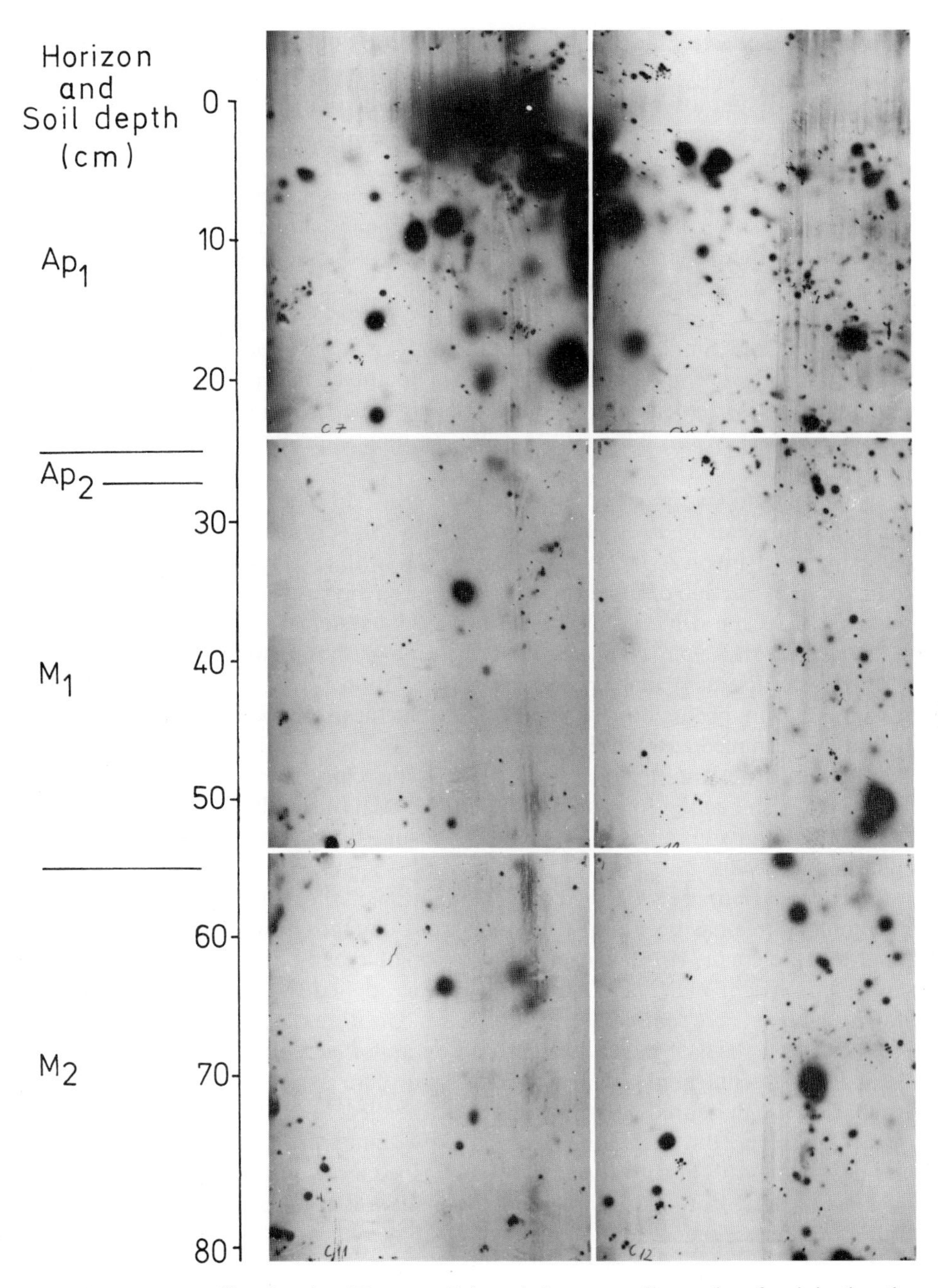

Fig. 8.3. Root profile of maize (*Zea mays* L.) made by autoradiography after injecting the plant stems with ^{32}P solution. (Mohr, 1975)

8.5.4 Root Studies with ^{14}C

A relative new technique is the use of the isotope ^{14}C in root ecology. It can be used for determining the total production and mortality rates of roots, the proportion of photosynthate of the plants which is incorporated in the root systems, the seasonality of photosynthate translocation by depth, and for distinguishing between live and dead roots.

In what is now regarded as a classical study Shamoot et al. (1968) determined the rhizo-deposition of organic debris in soil. They grew annual and perennial plant species in airtight chambers, allowing the plants to photosynthetically fix ^{14}C from $^{14}CO_2$. After the growing period of about seven months, the plants were harvested, roots removed, and the remaining root-derived soil organic matter analysed for ^{14}C. Similar studies to obtain information about the primary root production and decomposition in terrestrial plant communities were done by Dahlmann (1968), Sauerbeck and Johnen (1976), Sauerbeck et al. (1976), Singh and Coleman (1977), Wallace et al. (1977), Warembourg and Paul (1977), and Webb (1977).

Caldwell and Camp (1974) and Caldwell et al. (1977) determined the annual fractional turnover of root systems by using a $^{14}C/^{12}C$ dilution technique. This technique also involves labelling a plant stand with $^{14}CO_2$ early in the season, followed by extraction of root samples within seven days and a determination of the relative $^{14}C/^{12}C$ ratio in cellulose tissues of the root systems. The same procedure is conducted in the same plant stand at the end of the growing season. Then a turnover coefficient is calculated based on the dilution of the $^{14}C/^{12}C$ ratio over the season. The root production is the product of the turnover coefficient and the root biomass determined early in the season. Main advantage of this technique is that the necessity of determining the proportion of live and dead roots can be eliminated.

^{14}C can also be used for studying the spatial relationship between root systems of different plants in mixed plant stands (Neilson, 1964; Litav and Harper, 1967; Ellern et al., 1970).

An approach to use ^{14}C for getting information on the real amount of live roots in a root sample has been done by Sator and Bommer (1971), Sator (1972), and Singh and Coleman (1973) using a modification of a method of Ueno et al. (1967). Under an airtight plastic chamber $^{14}CO_2$ was applied to the leaves of a plant stand. Later soil core samples were taken. The roots were washed free from the soil, mounted on a sheet of paper and autoradiographes were developed using X-ray film. The percentage of the live roots in the core sample was calculated by comparing the measured length of total and labelled roots. In ecological routine studies, however, this technique is uneconomic, and the distinction between live and dead roots still must be done by visual inspection (see Sect. 11.8).

Nevertheless, from all the tracer methods the use of ^{14}C seems to be the most promising for ecological root studies.

8.5.5 Critical Evaluation

Radioactive tracers can be used to obtain information on the uptake of nutrients from different soil layers. A special advantage is that they provide information on roots without separating them from the soil. The techniques by which the isotope is injected into the soil or the plants generally take less time and labour than the traditional field methods in which washing and cleaning of the roots are necessary. Expensive equipment will not be of such concern in a large comprehensive programme, where available labour may be the greatest restriction.

It certainly seems elegant to do root studies in the field with modern isotopic laboratory equipment, but there are several limitations and objections which must be kept in mind (Burton, 1957; Wiersum, 1967b).

The use of most radioactive tracers presupposes that the soil profile is not stony. Also crevices and cracks in a soil can make the use of isotopes for root studies unsuitable. The half-life of the adapted isotopes is relatively short, so the tracers are more adopted for root studies with annual plants.

Many research workers have assumed that a high correlation exists between the active roots in the soil and the amount of the specific isotope absorbed by them during a distinct period. This generally occurs but it cannot be generalized for other elements. Experiments by plant physiologists have shown that, for example, roots which easily absorb phosphorus may not do so at the same intensity with another element (Russell, 1970, 1971, 1977). Thus the term root activity should be used with care (Burton, 1957; van Lieshout, 1960).

The initial enthusiasm following the introduction of tracer methods in ecological root research has cooled down. Indeed, it is really difficult to see how these methods can be effectively used on a larger scale. Certainly, a huge amount of root data has been collected with radioactive tracers but it is not easy to relate the data obtained in one experiment with those from another (Page and Gerwitz, 1974).

Radioactive tracer methods are of great importance in root physiology where a large part of the progress is based on the use of isotopes. It seems unlikely, however, that the methods can become routine for ecological root studies under field conditions. The use of isotopes cannot replace the measurements of the roots themselves by direct study methods. The most valuable information with radioactive tracers has been obtained by research workers who accompanied a tracer method with a direct root-study method. As complementary to other measurements the radioactive tracer methods will certainly keep their place in root ecology.

9. Other Methods

9.1 Measuring Root-Pulling Strength

Holbert and Koehler (1924) found that the resistance of maize roots to vertical pull was correlated with resistance to lodging. To measure the force required to uproot the maize plants, simple root-pulling machines have been constructed (Snell, 1966; Craig, 1968; Ortman et al., 1968). In principle a sash cord or a leather strap is placed around the base of the stalk and by a hand-powered winch the force necessary to pull the stalk out of the soil with the main part of the roots is measured by a tensiometer.

Zuber (1968) cut each maize plant so as to leave about 60 cm of the stalk above the soil surface. A cylinder made of a special wire was placed over this stub and, by vertical force, the plant was removed from the soil. A hydraulic cylinder, mounted in an upright position on the front of a small garden tractor, gave the vertical pull. The amount of vertical pull was measured on a tensiometer. The number of pounds pressure required to remove a plant vertically from the soil was the measure for root-pulling strength.

According to Zuber (1968) the optimum period for pulling maize roots is any time from flowering until physiological maturity. To obtain differences in root-pulling strength between cultivars, experiments should be made on plots with relatively homogeneous soil material. The absolute force necessary to uproot a plant differs greatly with varying soil structure and even more with soil water content. The best time to work with this method is directly after a heavy rain or after an irrigation period.

This technique is certainly a simple and cheap method for obtaining quantitative data about the root-pulling strength between cultivars, but there have been too few investigations made to recommend it as a routine method. Especially investigations where data of root-pulling strength were compared with data obtained by other root-study methods are lacking.

King and Beard (1969) used a vertical and Dunn and Engel (1970) a horizontal force technique for measuring the adhesion between the soil contained in a piece of sod and the underlying soil after a period of some weeks. Fraser (1962) and Fraser and Gardiner (1967) describe a method of studying the factors affecting tree stability by measuring the forces required to pull trees over.

9.2 Measuring Root-Clump Weight

Thompson (1968) proposed measuring the lodging resistance of maize by estimating the size of the root clumps which have been removed from the soil.

First the stalks were removed with a field silage harvester, then the root clumps were dug with a standard potato digger. The best time for doing this is about one month after flowering, presupposed that the upper soil layers are well soaked by a preceding rain.

The root clumps, consisting of the base of the plants, the attached roots, and the adhering soil, were rated visually for size on a scale from 1 (small) to 5 (large).Instead of visual estimations, the size of the clumps can also be determined by weighing. Thompson (1968) found high correlations between ratings and root-clump weights.

The root-clump method seems efficient and useful or studying lodging, anchorage, and other related factors of plants. Comparative studies made by Zuber (1968) and Nass and Zuber (1971) show a good agreement between root-pulling strength and root-clump weight. Perhaps the method will serve in the future as a test method in the field to see if other root parameters which have been found for young plants correlate with mature root systems. Nass and Zuber (1971) found that total root weight, root volume, and weight of the nodal roots were significantly correlated with root clump weight of mature plants grown under field conditions; but further research on this subject is necessary.

9.3 Measuring Root Tensile Strength

Another way to obtain quantitative data on lodging of agricultural crops is to measure the tensile strength of single roots. For this purpose the roots to be studied are washed free from soil, stored for some hours in alcohol (10%) or formalin (5%), and then placed in a simple instrument for measuring root tensile strength.

The known instruments consist of two plates. One plate is permanently mounted on a stand, the other is movable. A root with a distinct length (about 5–10 cm) is fastened by means of rubber strips between the two plates, and the force necessary to tear the root is measured. Such simple instruments have been described in detail by Kokkonen (1927), Pavlychenko (1942), and Spahr (1960). The force for tearing the roots generally is measured in grams, and serves as the index of the tensile strength of roots.

Tensile strength is correlated with the diameter of the roots. The diameter of the roots must therefore be determined first, and the data of the tensile strength can only be compared for roots within the diameter classes. Also the decaying of the roots must be taken into consideration before measurement, so for comparative studies plants should always be in the same stage of development.

Spahr (1960), who used the method to test the lodging of barley plants, found clear differences in the root tensile strength between strong and weak cultivars. Kokkonen (1927) measured the tensile strength of rye roots to find out relations to hibernation.

The method has not found too much approval because it is relatively time-consuming. Nevertheless, information on mechanical properties of the roots is valuable for many aspects in root ecology, so researchers should be encouraged to work in this field. Especially the dynamics of the root tensile strength during the life of a plant species should be subject of further research. Fundamental work on this question has been done by Slavoňovský (1976) during the last years.

9.4 Root Measurements Using Soil Sections

A technique for making thin sections of undisturbed soil with roots in natural position and to examine these under the microscope was proposed by Lund and Beals (1965). Melhuish (1968) described a procedure for determining the complete spatial arrangement of roots, by analysing serial sections, while Melhuish and Lang (1968, 1971) used the density of intersections of roots with plane surfaces to infer volumetric rooting densities based upon geometrical probability.

Melhuish's method involves the stabilization of soil cores by vacuum impregnation with a thermosetting polyester resin. Then the plane surfaces are prepared by using a surface-grinding machine. On these surfaces the root analysis is made by measuring the root void diameters left in the soil after drying and shrinkage of the roots using visible and ultraviolet fluorescence photography. The probable root length per unit soil volume is calculated by assessing the number of roots intersecting a plane of known area. The theoretical mathematical deviation for calculating this root parameter from the plane surfaces is outlined by Lang and Melhuish (1970, 1972).

The method in principle is very attractive for obtaining basic information about root–soil interface but it is not well adapted as a routine method for measuring rooting densities in a soil volume. The impregnation of the soil and the analysis of a plane surface is very tedious and time-consuming when measuring a large number of samples (Baldwin et al., 1971; Rowse and Phillips, 1974). Specific problems of using microtome sections for root studies are outlined by Stephan (1971).

Using microtome sections prepared by a combined technique of freezing and impregnation, Schwaar (1971, 1972, 1973) investigated roots in a superficial layer of a fen. His studies show that it is possible to distinguish in microtome section the roots of different plant species. Precondition for such a distinction is a detailed knowledge of the anatomical structure of the roots of the plant species under study. If more descriptions of the anatomical features of the plant species (e.g., Schröder, 1926, 1952) are available, this section method can probably be helpful for studying concurrent effects of the roots of different plant species, and this also in a quantitative way. It is hoped that when the fundamental basic research material from Kutschera (Austria) on the anatomical features of most of our plant species has been published, new impulses will come to research in this field.

9.5 Root-Detecting Method

The in situ study of the dynamic growth of single roots in natural soil profiles can also be done without using a glass wall method. Stoddard (1935) first placed permanent marked bands on single roots as a means of identification to determine the life span of roots; but it was Ladefoged (1939) in Denmark who used this technique on a large scale for studying root development of forest trees. He dug into the soil until he found a growing root, placed a thin copper wire on this root as a reference point, and measured the length from this point to the root tip. A stake was put in the soil to mark the location of the root, and then the root was reburied. To observe the root and measure its speed of growth during a distinct period, the root was uncovered again. Certainly the soil around the root is disturbed by every reexcavation, but Lagefoged (1939) found only small deviations when investigating a large number of roots.

Recently this in situ method with periodic excavations was used by Riedacker (1974) for root studies on *Eucalyptus* trees. This repeated measuring of single roots is, however, very time-consuming. According to Riedacker (1974) four persons had to work one day to measure 30 single roots from ten trees.

9.6 Mesh Bag Method

A modified method for testing direct effects of soil characteristics, especially chemical properties of the soil, on root growth was proposed by Lund et al. (1970). This "implanted soil mass technique" is an adaptation of a method used by Hendrickson and Veihmeyer (1931) to study the extension of roots in dry soil.

Holes 10-cm in diameter are drilled at determined positions from individual plants to a depth of about 30 cm. Then loosely woven plastic bags of the same diameter and length are inserted into these holes. These bags are then filled nearly to the surface with the soil which is to be tested. The top layer is covered with soil from the surrounding field and the location is marked. After adequate time (e.g., four weeks), the bags are dug out and the roots which have grown into the bags are recovered by washing.

The method has many applications, but seems ideal for testing the effect of soil-incorporated herbicides on root growth. When immobile compounds have to be tested, there will be no residual effects because the treated soil is completely removed by digging the bags out of the soil. The method should be practiced on a larger scale in field experiments as it also permits root penetrations studies across boundaries between two dissimilar soils (Garner and Telfair, 1954).

9.7 Root Replacement Method

A simple method of investigating new roots grown at specified places on living woody roots was proposed by Lyford and Wilson (1966). Adventitious "replacement roots" (Wilson and Horsley, 1970) are induced on exposed lateral roots still attached to the trees, and are grown under controlled environmental conditions in shallow trays using soil or other rooting media.

The method was tested by Lyford and Wilson (1966) with red maple (*Acer rubrum* L.), red oak (*Quercus rubra* L.), and white pine (*Pinus strobus* L.), but doubtless it can be used on most other tree species. It seems that the replacement roots grow at a nearly normal rate and with a normal habit, so the method permits relatively inexpensive studies of the root systems of mature trees. A distinction between old and new replacement roots can be made visually by staining the exposed roots with dye (Kaufmann, 1968).

Safford (1976) working with yellow birch (*Betula alleghaniensis* Britton) used the replacement method with an inert medium (perlite) and assumed that every replacement root produced would have a nutrient content governed by the nutrient status of the parent tree. The first results obtained by him show that the nutrient content determined in replacement roots can be an indicator of the nutrient status of the whole tree. However, further refinement of the method is necessary. Probably in connection with leaf sampling, the nutrient status of replacement roots can provide a better understanding of the mineral nutrition of trees.

9.8 Root Investigations with Paper Chromatography

An interesting method for distinguishing roots from closely related plant species by paper chromatography was proposed by Chilvers (1972). Roots from two different *Eucalyptus* species were sampled, washed, cut into short sections, extracted for several days in ethanol (95%), and spotted on chromatography paper. The pattern of spots observed for each species was reasonably consistent. In this way the roots of the two plant species were identified and their distribution mapped.

It seems that in some cases the use of paper chromatography can be a reliable method to distinguish between roots from related plant species. Recent progress in identifying cultivars of potatoes and cereals with gel electrophoretic methods also indicates that this kind of method can probably be used to distinguish roots from different plant species or cultivars having similar morphological or anatomical features. It seems a field for future research which can open up new possibilities in the study of competing root systems.

9.9 Electrical Methods

First experiments to measure the length of root systems by electrical methods were made by Kampe (1929), but they failed completely. Greenham and Cole (1949) tried to distinguish live and dead roots by measuring the electrical conductivity of the tissue of roots by inserting a probe into the roots. Dead roots had a higher electrical resistance than living roots. Davis and Johnson (1970) and Johnson and Davis (1971) used electronic growth sensors in the field for recording the growth pattern of sugar beets. Batchelder and Bouldin (1972) utilized the electrical conductance properties of single roots for determining root elongation rates through soil layers of different physical and chemical properties. However, all these studies were model experiments, and the techniques used are not suitable for practical ecological field research.

Recently, Chloupek (1972a, 1972b, 1976a, 1976b, 1977) measured electrical capacitance of root systems by means of an impedance bridge. One of the impedance bridges was connected with the basal part of the plant, and the other was grounded by a subsurface electrode 10, 20, or 30 cm away from the plant at a fixed depth. As living plant tissue is measured with the electrical capacitance, this method can give an indication of the size of the root system.

By measuring the electrical capacitance, nothing can be said about the spatial distribution of the roots in a soil profile. However, the method can perhaps be of interest in cases where only the relative size of the living root system of a plant is to be indicated. Further experiments are necessary to compare the validity of the data obtained by measuring the electrical capacitance with that found by other study methods.

9.10 Determination of Growth Rings in Tree Roots

Counting growth rings in tree roots can be used to determine root age. In ecological research the determination of the number and size of growth rings in tree roots can give hints to the environmental conditions on root development. Research in this field has been done by Vater (1927), Wagenhoff (1938), Joachim (1953), Wilson (1964, 1975), Fayle (1968) and others. The techniques of determining growth rings in trees and many of the problems involved are outlined in the new standard book on dendrochronology by Fritts (1976).

9.11 Investigations of Root Hairs

Root hairs, the small narrow bulges on the epidermal cells of roots, having a diameter of about 5–15 μm, a length of 3–10 mm, and a normal life-span of few days, can be produced by nearly all plant species. Even most of the aquatic plants can have abundant root hairs (Shannon, 1953).

Root hairs enlarge the surface of the roots considerably, and serve as anchorage for the roots when they grow into the soil. Their significance as organs for nutrient and water uptake is controversial (Russell, 1970, 1971, 1977). Studies of root hairs are primarily in the field of root physiology.

Root hair studies in non-soil substrates are made predominantly in humid air where the optimum hair production occurs (Schwarz, 1883), on wet filter paper (Snow, 1905), or in sand or water culture (Bergmann, 1958). For studies of the development and decaying of root hairs, glass wall methods can give good results (Richardson, 1953).

In field studies sampling methods will be used where the roots are freed from the soil by washing, and the root hairs are determined for distinct root segments. The soil-root samples are washed in running water without using cutting or scraping tools to speed the washing process. After all the soil has been washed away, the root system is placed in alcohol so it is preserved for the additional measurements. For the root hair study, segments of individual roots are placed in Petri dishes containing water. Then the hairs are counted on representative units of the root segments by means of a binocular or a microscope. Also diameter and length of the hairs can be measured and calculations of their surface area can be made. This technique was used by Dittmer (1937, 1938, 1949).

For better observation, the root hairs can be stained by dipping the root segments in a solution of methylene blue for a few minutes (Cormack, 1944). Mc Elgunn and Harrison (1969) distinguished living root hairs from dead ones by staining them with 1% solution of neutral red. Hairs with living cells take up neutral red and retain it, whereas the stain will not stay on dead cells.

It is astonishing how tightly the living root hairs cling to the roots. The objection that by washing the roots all the root hairs are lost is not valid. Certainly, they are not so ideally arranged after the washing process as in the soil where they have grown; but even in such disarrangement, their number and size can be counted, or at least estimated (Bole, 1973). Most of the root-hair studies on roots grown in soil have been made on cereals and other grasses, in a few cases also on tree roots (Kolesnikov, 1971).

9.12 Determination of Root Nodules

One important task in root research is to determine root nodules, the symbiotic, nitrogen-fixing bacteria tubercle on root systems of legumes and some non-leguminous plants. The root ecologist generally is interested in the number of nodules. After the root sample has been washed and cleaned (see Sect. 11) the number of nodules is determined and calculated on a per plant or a root length basis. A recommended time for determining nodules is when the plants are shortly before bloom because then the nodules are plump, healthy and flesh-coloured. For studying the dynamics of root nodule development glass wall methods are to be recommended.

The isolation of the bacteria of the root nodules, determination of their species and other histological investigations belong to the field of microbiology (Fred et al., 1932; Mishustin and Shil'nikova, 1971; Becking, 1975; Dart, 1975).

9.13 Determination of Mycorrhizae

The root ecologist is confronted with ectomycorrhizae when studying root systems of species in Pinaceae, Betulaceae, Fagaceae, and a few other plant families. Here the fungus forms a compact layer over the root surface and produces intercellular hyphae in the cortex. To define the ectomycorrhizae on the roots macroscopic and microscopic characters can be used such as morphological features, colour, chemical colour reaction, and ultraviolet light fluorescence. A practical widely approved key for identification of ectomycorrhizae has been approved by Dominik (1969). Further methodological hints are outlined by Marks and Kozlowski (1973).

The most common type of mycorrhizae is the VA-mycorrhizae, found on most of the agricultural and horticultural crops, as well as on many forest trees. It consists of fungi that form vesicles, arbuscules, and non-septate hyphae in root cortices, and a loose mycelium in the soil around infected roots. This kind of mycorrhizal infection produces only very little change in root morphology, and is often overlooked.

Techniques to diagnose and quantify VA-mycorrhizae infection are lacking. The simplest methods are clearing and staining procedures, that of Phillips and Hayman (1970) allowing rapid detection of the mycorrhizae by microscopic examination. Recently Ambler and Young (1977) proposed a quantitative technique where combined measurements of the total root length and the root length infected with VA-mycorrhizae can be made. Steps include separating roots from a known soil volume, cutting these roots into short segments, clearing tissue in KOH (10%), staining fungi with trypan blue, subsampling root segments in glycerol, and counting the segments with the line-intersection method using a microscope. With this technique also other root features, for example, root hairs, tips, diameters, and condition of the cortex can be studied.

Mycorrhizae research has developed to a special field of research, mainly done by microbiologists (Harley, 1969; Mosse, 1973; Gerdemann, 1974, 1975).

10. Container Methods

10.1 General Features

Container-grown root systems are primarily used for studying the morphology, physiology, biochemistry, and to some extent also the ecology of root systems. The container approach allows the researchers to isolate individual environmental factors which have interacting influence on root growth in natural soil profiles. By utilizing different combinations of environmental factors a great deal of basic information can be obtained regarding the significance of single environmental factors and the interactions between these factors on root growth.

A major advantage of the container methods rests in the fact that container-grown plants are generally easier to handle and study than are field-grown plants. Another advantage is that the growing conditions can be uniformly replicated many times. Major disadvantages of the container methods center around the unnatural conditions, such as use of disturbed soil or artificial rooting media, and possible lack of room for normal root spread and distribution. The lack of root competition from other plants may also be significant, as may be the presence or absence of specific soil organisms.

Container-grown root systems have been studied extensively since the end of the 19th century and an extensive literature has developed. The procedures utilized for growing plants in containers have been described in detail in handbooks and manuals such as Schropp (1951), Giesecke (1954), Hewitt (1966) and others. Therefore this section will only outline and comment on those variations of methods which are adapted for studying root systems.

10.2 Rooting Volume and Container Size

The size and shape of the container places a finite limit on the rooting volume both in terms of horizontal spread and depth of penetration. This may be a serious problem when dealing with plants having extensive root systems. The size of the container controls the amount of soil available to the root system and may considerably influence the development of the plant's root system and aboveground parts. Although this fact was convincingly demonstrated by Hellriegel (1883), it took over 50 years before researchers began to consider the effect of the size and shape of the container on their results.

The restriction of the unlimited horizontal and vertical root development may partially be overcome by studying only the early stages of plant development or by studying plants with small root systems. Typical examples of root investigations involving seedling stages are those by Aubertin and Kardos (1965a, 1965b), Hentrich (1966), Brown (1969), Eavis (1972), and Kar and Varade (1972).

When larger root systems are to be studied, the nature of the root system and study objectives will determine if the container should have a large diameter or a large vertical rooting volume or both. According to Hocking and Mitchell (1975) where increased rooting volume is needed for special cases, it is better to use containers with a relatively large diameter. However, most researchers who study container-grown root systems recommend that the container have an adequate rooting volume in the vertical direction. This is in part due to a root's tendency to turn and grow downwards upon being restricted by the walls of its container. As a result, long vertical boxes and tubes of relatively small to moderate diameter generally are the predominant type used in ecological root studies.

The size of the container and rooting volume influence the nutrient and water supply available to the plant. Plant growth will be adversely affected if this supply is limited or present in excess amounts. This means that the nutrient and water supply must be regulated individually according to the container size and the existent rooting volume. Special recommendations and hints are outlined in Cook and Millar (1946), Sokolow (1956), Stevenson (1967), Cornforth (1968), and Grosse-Brauckmann (1972).

10.3 Types of Containers

10.3.1 Small Pots

The simplest kinds of containers for root studies are ordinary clay flower pots, Mitscherlich pots, and different types of round or square plastic pots. Also quite useful and inexpensive are cylindrical cardboard cartons impregnated or coated with wax (Giskin and Kohnke, 1965). Small transparent containers can be placed in larger opaque ones, the transparent containers only being taken out for observation of the roots (Fig. 10.1).

The shallow depth and small volume of most of the small pots limit root spread to such an extent that in most cases, even with young plants, the roots will be concentrated near the walls and around the bottom of the pot. Pot material will influence the root concentration to a large extent. In porous clay pots, roots tend to concentrate outside the soil mass between the soil mass and the clay wall. In contrast, with non-porous plastic pots, roots tend to be more ramified throughout the soil mass with a smaller proportion next to the pot wall.

According to Jones and Haskins (1935) and Kutschera-Mitter (1971) the reason for this phenomenon is moisture movement. The walls of the porous clay pots absorb moisture from the soil and evaporate the moisture into the outside air. As the distribution of roots is closely tied to soil moisture movement, a massive root concentration occurs at the wall. Nutrients in solution tend to follow the movement of soil moisture. Thus with time the nutrient content will be found to decrease in concentration from the centre and increase outwards towards the clay wall.

Material also influences the temperature of the soil in the pots. In comparing clay and plastic pots Bunt and Kulwiec (1970) found that the soil in clay pots was cooler than in plastic pots. The temperature difference was attributed to the cooling

Fig. 10.1. Root studies with young cereal plants grown in transparent plastic pots. For protecting the roots from light the plastic pots are inserted in wax-impregnated cardboard pots of similar size

effect of water evaporating through the porous clay walls. Temperature differences between the two types of pots ranged from about 1° C at night in the winter to about 4° C in the summer. According to Rupprecht (1954) the daily variation in soil temperatures are greater near the walls of the pots.

These differences may have little or no significance when small pots are used to grow plants for comparative root studies such as to test different liming treatments (Longenecker and Merkle, 1952; Wind, 1967), to investigate the influence of herbicides on root development (Hartmann, 1969), or similar studies. However, it should be kept in mind that while such studies can rapidly provide information regarding a specific research problem, the smaller the pot size and the longer the plants remain in the pot, the more limited will be the general validity of the results obtained.

10.3.2 Boxes and Tubes

Large-scale ecological root studies have been accomplished in long boxes and tubes of various designs. Some of the earlier containers were heavy galvanized iron boxes 30 × 30 cm square and up to 100 cm deep (Weaver and Himmel, 1929, 1930; Roemer, 1932; Partridge, 1940). Many of these metal boxes had removable front sides. Other boxes up to about 200 cm long with removable doors were made of wood (Sekera, 1928; Görbing, 1930; Goedewaagen, 1933, 1937; Bär and Tseretheli, 1943; Weaver and Zink, 1945; Wadleigh et al., 1947; Szembek, 1957; Oberländer and Zeller, 1964; Schuurman and Goedewaagen, 1971; Onderdonk and Ketchenson, 1973a). Some research workers impregnated the wooden boxes with

Fig. 10.2. Tree root studies in PVC tubes at the Institute for Soil Fertility in Haren-Groningen (Netherlands)

paraffin. Flocker and Timm (1969) used acrylic plastic containers 5 × 11 × 80 cm in their study.

Cylindrical tubes of different material and dimensions have been used more often than boxes. Large iron cylinders with sloping bottom (Weaver et al., 1924), aluminium cylinders (Hanson and Juska, 1961), porous drain pipes (Jacques, 1937b), and large clay cylinders with a diameter of 32 cm and a length of 60 cm (Gliemeroth, 1953a, 1957) have been used at various times. Concrete cylinders have not been widely used because they are too heavy and difficult to handle (Frankena and Goedewaagen, 1942). Asbestos tubes with an internal diameter of 15 cm and a height of 75 cm have been successfully used, primarily by researchers in the Netherlands (Schuurman, 1965, 1971 a, 1971 b; Wiersum, 1967 a; Schuurman and De Bor, 1970, 1974; Schuurman and Goedewaagen, 1971; Schuurman and Knot, 1974). These tubes can be relatively easily cut longitudinally thus facilitating washing. Cut tubes are reusable. The two halves should be held together with metal straps and the seams sealed to prevent loss of water (Ellern, 1968).

In recent years, plastic cylinders have been used with good success (Schuurman and Goedewaagen, 1971; Reicosky et al., 1972). Before the tubes are filled with soil their bottoms are closed with porous nylon or wire mesh. The tubes are placed in shallow dishes with a constant water table (Fig. 10.2).

Frequently a temperature problem exists when using boxes and tubes. If the boxes or tubes are left in the greenhouse, the temperature of the soil within the box or tube will be at the temperature of the greenhouse and quite probably will be substantially different from that in the natural undisturbed soil where the roots would normally grow. Since root development is influenced by temperature, many researchers have inserted their containers into the ground (Sekera, 1928; Jacques, 1937b; Hanson and Juska, 1961; Schuurman and Goedewaagen, 1971, and others). For many research questions the discrepancy in soil temperature may not be of great importance. But by inserting the containers in the soil, the environmental conditions experienced by the roots are closer to those in a natural soil profile. Some researchers have installed permanent cement or concrete boxes in the soil to overcome this temperature problem (Schulze, 1906, 1911/14; Obermayer, 1939).

10.3.3 Glass-Faced Containers

Most container-grown root studies utilize glass-faced boxes and tubes, generally inclined at angles of about 3°–25° from the vertical (glass side down) to make phenological observation of root growth and development.

The majority of glass-faced boxes are made of wood or metal with one of the side walls containing a removable glass (or Plexiglas) plate. A cover plate of opaque material, such as aluminium foil, is used to cover the glass and protect the roots from light (Fig. 10.3). The sizes of the boxes vary, depending on the species under observation and on the research aim. The sides and the bottom of the boxes are usually sealed to prevent water and nutrient loss through leakage. Frequently the boxes will be fitted with a drain so that they may serve as a lysimeter.

The techniques of recording the roots behind the glass are outlined in Section 7.3.3. At the end of an experiment, the glass plate can be removed and a thin layer of about 5–10 mm soil can be washed away. More details of the root system of the plant under study can thus be observed than by only looking through the glass (Fig. 10.4).

Numerous variations on glass-faced boxes have been developed. Triangular boxes with glass plates on all three sides have been used with success by Fröhlich and Dietze (1970). Special vertical standing root study boxes 6.5 × 120 × 120 cm with glass observation surfaces on both sides, which were mounted on trolleys, are described by Werenfels (1967), Schumacher et al. (1971), and de Haas and Hein (1973). Permanent, free-standing boxes with vertical glass plates at the front can be located in a greenhouse for special research situations (Fig. 10.5). Examples where glass-faced boxes have been inserted in the soil under field conditions are few (e.g., Engler, 1903; Lavin, 1961). A modern root-observation-chamber field installation with 24 steel chambers having glass walls on two of its four sides and working as a weighable lysimeter station has been constructed in Texas, USA (Arkin et al., 1978).

Numerous investigations, utilizing a variety of plant species, have been conducted since the German botanist Sachs (1873) first described the construction

Fig. 10.3. Glass-faced boxes placed in an inclined position. The glass plates in the front are covered with aluminium foil which is removed for observation of the roots

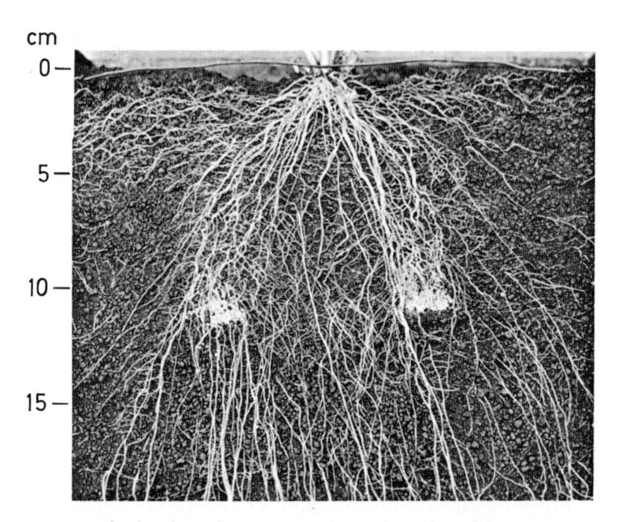

Fig. 10.4. Root system of a barley plant grown in a glass-faced box after removing the glass panel and washing away a thin layer of soil. The two white spots in the centre are grains of phosphorus fertilizer. (Böhm, 1974a)

and use of glass-faced boxes. The reader is referred to: Engler, 1903; Bodo, 1926; Dean, 1929; Ostermann, 1931; Kauter, 1933; Tukey and Brase, 1938; Rogers, 1939c; Johanson and Muzik, 1961; Howland and Griffith, 1961; Lavin, 1961; Larson, 1962; Muzik and Whitworth, 1962; Leibundgut et al., 1963; Hurd, 1964, 1968, 1974; Salim et al., 1965; Asher and Ozanne, 1966; Straub, 1966; van der Post

Fig. 10.5. Glass-faced compartments in a greenhouse at the Tobacco Research Institute at Forchheim (Western Germany)

and van der Meijs, 1968; Wiese, 1968; Larson and Schubert, 1969; McElgunn and Lawrence, 1970; Dunn and Engel, 1971; Mosher and Miller, 1972; Bockemühl, 1973; Evetts and Burnside, 1973; Pearson et al., 1973; Böhm, 1974a; Hurd and Spratt, 1975; Shaver and Billings, 1975; Billings et al., 1976, and Parao et al., 1976.

For root-penetration studies, small glass tubes about 5 cm in diameter and up to 120 cm long have proved satisfactory (Kittock and Patterson, 1959). Bilan (1964), Newmann (1966b), Voorhees (1976) and Taylor et al. (1978) used tubes made of Plexiglas. Murdoch et al. (1974) developed PVC tubes with removable observation windows made of transparent acetate plastic. Nilsson (1965, 1969, 1973) devised a special equipment by hanging transparent tubes about 6.5 cm in diameter and one meter length from a scaffold which maintained the tubes in an inclined position (Fig. 10.6). The tubes, which contain an artificial substrate such as perlite, are supplied automatically with nutrient solution. This kind of equipment which permits also independent control of the temperature of the roots and the shoots by hanging the tubes into a temperature-regulated water bath, has been used predominantly for studying problems of root pathology, but it can also be used successfully by plant breeders for screening cultivars (Mac Key, 1973).

Fig. 10.6. View of the Nilsson tube-culture equipment. (With kind permission of J. Mac Key, Uppsala, Sweden)

Small root observation boxes have been described by Gerlagh (1966) and Estey (1968, 1970) which may be utilized for microscopic examinations of roots and root diseases. Schmidt (1934) used small glass dishes in which he evaluated the germination power of seeds by estimating their rooting intensity at an early stage.

The same limitations as in the use of the glass wall methods in the field (see Sect. 7) also apply to glass-faced containers used in the laboratory or greenhouse. The problem of root concentration will probably be greater in the containers due to the limited rooting volume. Even so, the glass-faced or glass wall methods are unsurpassed for phenological root studies.

10.3.4 Flexible Tubes

A recent variation on the glass tube method involves using transparent or translucent tubes made of flexible plastic film (Kausch, 1967; Derera et al., 1969; Böhm, 1972, 1973b). Common film tubes range from 6–30 cm in diameter and from 60–150 cm in length. Some problems may be encountered in supporting the larger film tubes and in preventing stretching and possible rupture of the film due to the mass of material. The film thickness should be about 0.2 mm. The film tubes can be manufactured as needed by the researcher from rolls of film or be purchased ready to use. In most cases it is more economical and practical for the researcher to form his own film tubes by sealing the edge and bottom of a folded sheet of film of appropriate size.

Fig. 10.7. Flexible polyethylene tubes supported in wooden stands. For protecting the roots from light, the tubes are wrapped with aluminium foil

Details for using polyethylene tubes 100 cm long and 15 cm in diameter have been outlined by Böhm (1972). He used premanufactured tubes, closed the bottom with plastic sheet, and perforated this base with 100 needle holes to allow water to penetrate. The tubes were filled with sieved soil which was dampened during filling. The tops of these soil columns were held circular by means of strong PVC rings 4 cm high. The filled tubes were placed in small dishes, surrounded with aluminium foil and stabilized in a wooden stand (Fig. 10.7). The plants growing in these film tubes were watered by a constant water supply maintained in the dishes and from above as needed.

Root growth information may be obtained from these columns in two ways. The entire root mass may be washed out of the tubes intact (see Sect. 10.8) or the soil column may be placed in an appropriate wooden box or tray and the column cut into segments without removing the plastic film (Fig. 10.8). Roots in a specific segment can then be separated from the rooting medium by washing. This allows a precise determination of rooting density and characteristics at specific depths.

The larger the tube, the more the problem of uniformly packing the soil. A major problem is the development of folds or wrinkles in the lower part of the tube. These folds will often serve as pathways for roots and may tend to invalidate the results obtained. Folds generally develop when loose soil is placed into inadequately supported tubes. The tendency is for the whole column (tube and soil) to settle rather than for the soil to settle within the tube. This problem can be minimized by placing the film tube in a proper-sized supporting packing frame and by using sand or rooting media other than soil.

The advantages of using film tubes are considerable. Expenditure of time and effort in freeing the roots from the rooting media may be reduced up to one fourth

Fig. 10.8. A soil-root segment easily obtained by cutting a foil-wrapped polyethylene tube with a knife. (Böhm, 1972)

as compared to the use of rigid cylinders. The cost of film tubes is 10–20 times cheaper than rigid glass or plastic cylinders. However, since the film is cut for the washing process, each tube can be used only once.

10.4 Rooting Media

The development of a plant depends on the environmental conditions of the culture medium in which its root system is growing. Generally, changes in culture medium influence the roots more than the shoot. For example, the distribution and extent of a root system may be drastically modified by the presence of a strong, resistant layer even though the top portion of the plant shows little or no sign of influence.

A comparative study using wheat plants and various culture media was conducted by Mac Key (1973). The results, shown in Figure 10.9, effectively demonstrate the differences in root and shoot development due to different culture media. Soil offers the best balance between shoot and root production. The rooting habit of the soil-grown plant is similar to that of plants grown in natural soil profiles under field conditions. The "normal" root systems from the soil contrast sharply with the proliferous, stout and little-branched root systems grown in water culture. Roots grown in sand culture or well-irrigated sand generally develop root systems similar to that developed in water culture. Perlite (a highly water-absorbing, porous, pumice material) also produces root system development similar to that in water culture. Vermiculite (a hydrophilic mica substrate) promotes "normal" root growth, while polytherm (a hydrophobic, ball- shaped, light plastic

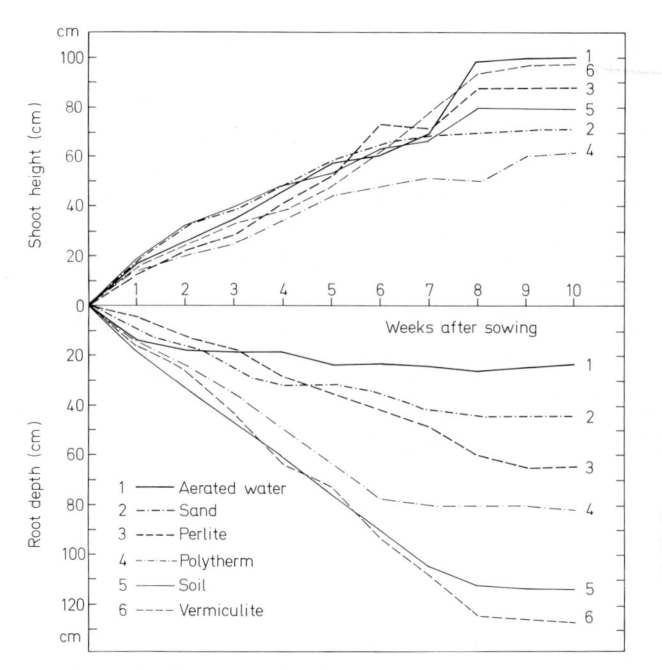

Fig. 10.9. Shoot and root development of spring wheat grown in six different culture media. (Mac Key, 1973)

material) causes an intermediate reaction. These results should not be generalized since they were obtained under optimum nutrient and moisture conditions. Under stress conditions the differences may vary. The comparative study does however reveal that one must interpret the data very carefully whenever non-soil substrates are used. A similar study on the influence of rooting media on root development and root structure of Douglas fir cuttings was made by Copes (1977).

The ability of roots to penetrate substrates of different rigidity has been studied using paraffin or wax layers of different hardness (Hendrickson and Veihmeyer, 1931; Hunter and Kelley, 1946b; Mederski and Wilson, 1960; Taylor and Gardner, 1960a, 1960b; Stone and Mulkey, 1969; Thorup, 1969, and others). Such non-porous, uniform substrates can have advantages over soil or other porous rooting media in obtaining causal information as to a root's penetrating abilities. But such studies, while highly informative, cannot totally replace studies in real soil substrate.

10.5 Filling the Containers

How the containers are filled may have a substantial impact on the outcome of a root-growth study. Layering or separation of soil particles according to size during the filling process or difference in bulk densities must be avoided. Packing soil columns to a predeterminated bulk density is an art. Whenever possible the same person should pack all soil columns for a study using exactly the same procedure for each column.

In general, the soil which is to serve as the rooting medium in the containers should first be sieved through suitably sized mesh, depending on the kind of soil and experimental objective. Best sieving results are obtained using nearly air-dry soil. Appropriate amounts of fertilizers should be added to the sieved soil in accordance with the study objective.

The containers should not be filled with dry soil as the soil particles may separate according to particle size during the filling process. To overcome this separation problem dry soil should be moistened with sufficient water to produce a friable condition, and mixed thoroughly.

A common procedure for filling a container is to place a layer of unstratified soil, about 5–10 cm deep, into the container and to stamp it down by means of a simple wooden or metal stamper. In rigid containers the soil along the edges can be pressed down with a narrow curved iron bar. After stamping the first soil portion, the topmost 2–5 mm of the compacted layer is loosened again to prevent the formation of an interface when the next addition of soil is placed into the container. The entire process of adding thin layers of soil, stamping, and loosening the soil is repeated until the container is full.

Reicosky et al. (1972) recommend filling large soil columns by gently tapping and rotating the column as the soil is poured into the column. Others recommend lifting the container and lightly dropping it on a hard surface to settle and firm the soil within the container.

The deeper the container the more difficult it is to obtain a uniform bulk density from top to bottom. Only care and experience will allow the researcher consistently to obtain uniformly packed columns with the same bulk density.

If it is not necessary to obtain a predetermined bulk density, it may be advantageous simply to fill the container with sieved soil without stamping, tapping or jarring. The soil can then be settled and firmed up through wetting and draining of the soil column (Bilan, 1964; Taylor et al., 1978). The wetting and drainage is repeated until no significant settling of the soil is observed. Additional soil should be added to the soil within the container to bring the level of soil within the container to the desired level. There is some possibility in this method for particle migration and the development of slightly denser layers.

Settling of the soils in any container where the soil has not been packed to a high bulk density may occur with time and watering. It is therefore recommended that all filled containers be watered several times and allowed to stand for several days or longer before beginning the experiment. It is especially important to allow for settling when substantial depths of soil are involved. When a large rooting volume in the vertical direction is to be used, it is often advantageous to allow several months of watering and settling time between filling and the start of the experiment.

Numerous container experiments have been conducted with the objective of evaluating the influence of bulk density or soil strength on root growth. Small metal tubes were filled with soil and compacted by a hammer, piston or by a special compression apparatus to the desired volume and strength. The reader is referred to the following selected publications for detailed methodology: Veihmeyer and Hendrickson (1948); Meredith and Patrick (1961); Trouse and Humbert (1961);

Zimmermann and Kardos (1961); Taylor and Gardner (1963); Barley et al. (1965); Stolzy and Barley (1968). The technique of producing soil columns having various soil layers with different bulk densities in long asbestos tubes has been described by Schuurman (1965)

10.6 Seed Technique

Generally the seed is placed into the containers without pre-germination. Care should be exercised to use uniform-sized seeds. Experiments with cereals by Heinisch (1938) and Michael and Bergmann (1954) indicate that the larger the grain, the greater is the number and the length of the primary roots. Evidence seems to indicate that non-genetic differences in seed size appear to influence shoot and root growth equally (Mac Key, 1973).

Aubertin and Kardos (1964) proposed a technique to utilize pre-germinated seedlings having specified root lengths in their short-term root penetration studies. Where woody species, such as forest or orchard trees, are involved, it is often advisable to transplant container-grown seedlings, with all the surrounding soil, from the small starter containers into larger ones (Parker, 1949).

10.7 Irrigation Problems

In most container root studies surface watering is utilized based on appearance and judgment. Adequate water is added to each pot on a regular basis so that the plants always have an adequate amount of moisture for optimum growth. Automatic watering, which seeps, drips, or periodically sprays water on the plants and/or soil, is becoming more and more common. Frequently a layer of gravel or coarse sand is placed on the top of the soil surface to absorb the force of the applied water and thereby prevent damage to the soil.

Containers may also be watered from below. This method has several advantages. Good aeration, especially in the upper portion of the soil column, is enhanced, air entrapment does not occur, soil structure is maintained, nutrient loss through the drainage holes is reduced, and the time and labour required for surface irrigation can be saved. In practice, the bottoms of the containers are placed in shallow dishes in which a constant level of water is maintained (Baumann, 1948; Baumann and Klauss, 1955; Wiersum, 1967a; Schuurman and Goedewaagen, 1971, see also Figs. 10.2 and 10.7). To prevent evaporation from these dishes they may be covered with a rubber collar or a fitted plastic film.

In some cases a combination of watering from below and from the surface should be utilized. This is especially necessary where long columns are utilized. The capillary and adsorptive forces of the soil matrix may not be able to draw sufficient water into the root zone to maintain adequate moisture for the plant, especially if the plant's root system is concentrated near the surface as in the early stages of growth, or if coarse-textured rooting media are used.

Special soil–plant–water relations studies require more control of the water regime than can be provided by the above procedures. Many root study

Fig. 10.10. Washing complete root systems out of a flexible polyethylene tube by the use of an oscillating sprinkler. (Böhm, 1972)

experiments involve subjecting the plant to moisture stresses or a determination of actual transpiration usage. Tensiometers or moisture blocks placed at various depths in the container can give continuous in situ information about the soil moisture status at specified locations within the column (Davis, 1940; Hunter and Kelley, 1946a; Werenfels, 1967; Cullen et al., 1972; for detailed methodological hints see Slavík, 1974).

10.8 Special Washing Procedures

In contrast to most field methods, the container methods allow for a relatively easy recovery of the entire root system of the plants. Roots grown in small containers, like Mitscherlich pots, are soaked with water to about field capacity, then placed on an inclined washing table. Starting from the surface, the soil is washed away by means of a hand sprinkler. Another possibility is to put the complete soil-root mass out of the container into a pail to be suspended and washed as described in Sections 11.4 to 11.6.

Long boxes with removable front sides are placed in an inclined position (not more then 20°) and the soil is washed away by means of a hand sprinkler. General washing instructions are the same as outlined for washing out needleboard monoliths taken from natural soil profiles (see Sect. 4.6.4).

In a similar manner, the complete root system of plants grown in polyethylene tubes can also be washed out. The additional use of an oscillating sprinkler (Fig. 10.10) facilitates the washing procedure. Washing out rigid tubes takes considerably more time. It also starts from the bottom of the tubes which are placed

Fig. 10.11. Washed root system of an oil radish plant (*Raphanus sativus* L.) grown in a 90-cm long flexible polyethylene tube

in an inclined position. Schuurman and Goedewaagen (1971) recommend using a special sprinkler of adequate length which allows either sprinkling frontally or sideways, regulated by a two-way valve.

For photographing, the washed root systems are placed in flat black dishes or basins where remaining soil particles and other impurities are removed under water. The cleaned root system later can be re-arranged by tweezers and needles and photographed (Fig. 10.11). Hints for photographing washed root systems are given in Section 4.6.5.

10.9 Modified Container Methods

10.9.1 Cage Method

The cage method introduced by King (1892) for root studies in undisturbed natural soil profiles (see Sect. 4.5) was also used in a modified manner by the same research worker and by Hays (1893) for root studies in self-constructed "containers". Simple round or square frames made of wood or metal of different

Fig. 10.12. Principle of the cage method. Root system of a young soybean plant grown in a plastic box with screen bottoms. (With kind permission of H.M. Taylor, Ames, Iowa, USA)

sizes were placed in appropriate sized holes in the field. Sieved soil was then placed into the hole and stamped down with a stamper to form a layer of soil about 5 cm thick. A 5-cm mesh poultry netting or other galvanized wire netting was laid over this soil layer and attached to the frame. The procedure of adding soil and placing wire netting over the soil layer at 5-cm intervals was repeated as often as necessary to fill the hole completely. This resulted in a sunken "container" having wire supports about 5 cm apart. Subsequently when plants were grown in these sunken containers, their roots grew within and between the supporting wire. Upon completion of the growth period, the netted frames were freed from the surrounding soil, transported to a washing place, and starting from the top, the root system was freed using a sprinkler. The wire netting provided support to the root system and kept the individual roots in essentially the same position they occupied in the undisturbed soil. This procedure provides a three-dimensional representation of the root system as it existed in the soil.

Later several researchers constructed rigid boxes with one or more detachable side walls. Wire netting was placed horizontally at different depths within the boxes (Fig. 10.12). Intensive studies with this type of cage container were made by Venkatraman and Thomas (1924), Lee (1926), Ostermann (1931), and Bergmann (1954). More recently Pittman (1962) used the cage method with success to study the influence of magnetotropism on the growth direction of roots.

The cage method seems well adapted for studies designed to observe tropism responses by roots during the early stage of plant development. Due to the support of the wire netting the direction of root growth is maintained in the exposed root system. The use of sandy soil or an artificial substrate such as glass beads facilitates the washing process. Even so, in general, the same limitations as outlined in Section 4.5 apply to this method, for example, that by the washing process the finer roots stick to the thicker ones and so often only a distorted root system will be obtained.

10.9.2 Needleboard Method

In this method plants are usually grown in wooden or metal boxes of various sizes; one side wall is always detachable. When the root system is to be examined, the removable side wall is replaced by a needleboard (see Sect. 4.6). This board is pressed against the soil until the needles are completely inserted. Then the whole box is turned upside down with the needleboard lying underneath and the box is carefully removed. Finally, the soil monolith now lying on the needleboard will be washed away until the root system is freed as described in Section 4.6.4.

The use of needleboards in root box studies was proposed and used on a large scale by Rotmistrov (1909) in Russia. His instructive root pictures thus obtained influenced several workers to use this study method (Polle, 1910; Maschhaupt, 1915; Goedewaagen, 1933, 1955; Pinthus and Eshel, 1962; Schuurman and Goedewaagen, 1971). Replacing the detachable side wall with a needleboard wall at the beginning of the experiment (Hess, 1949) is not recommended, as the needles will hinder the filling of the boxes with the soil. Only if soil is filled in the boxes and packed by watering will the needles not interfere with filling. The advantages and limitations of this method are outlined in Section 4.6.7.

10.9.3 Root Training by Plastic Tubes

A simple technique for directing roots through narrow plastic tubes to a distinct soil depth at which they fan out from the lower end of the tubes has been described by De Roo and Wiersum (1963), who filled rigid plastic tubes of different lengths and inside diameters of 2.5 or 4 cm with a soil-peat-sand mixture. The lower end of the tubes was closed with a layer of cheesecloth by means of a rubber band. Then seeds or young seedlings of various annual crops were planted in these tubes. When the roots appeared at the bottom of the tubes, these were transferred into wooden boxes partially filled with soil. After the cheesecloth had been removed, the tubes with the plants were set upright into the box by inserting them some centimeters into the soil. After adequate time (1–2 months) the detachable front side of the boxes was replaced by a needleboard, and the root systems developed in these boxes were washed out.

These tubes with young plants grown in them can also be inserted in the field in auger holes vertical to a distinct soil layer in which the growth of the roots is to be studied. De Roo (1966b) demonstrated the usefulness of this technique in an examination of whether roots, once introduced through a compact plough pan, would grow in the horizon beneath the pan as well as in the upper soil.

With this tube technique, used either in greenhouse or field experiments, possible differences in the ability of roots of various plant species to grow upwards in friable soil can be tested. Indications about how far root systems will return from deep soil layers upwards can also be obtained by growing plants in U-shaped tubes (De Roo, 1966a).

10.9.4 Split-Root Technique

The split-root technique allows the researcher to divide the root system of a young plant and to place portions of it into separate containers having rooting media reflecting different treatments. It has predominantly been used to study the influence of different fertilizer treatments on root growth (Gile and Carrero, 1917; Sideris, 1932; Duncan and Ohlrogge, 1958; Wiersum, 1958; Rios and Pearson, 1964; Boatwright and Ferguson, 1967, and others). The preferred culture medium for split-root studies is water. The split-root technique is very well adapted to the field of plant nutrition and is widely utilized by researchers in this discipline.

Examples of the split-root technique of growing divided root systems in different soil containers are given by Thomas (1927), Giskin and Kohnke (1965), Whitcomb et al. (1969), and Whitcomb (1972).

10.9.5 Undisturbed Soil Monoliths

The idea of utilizing undisturbed soils in containers within the laboratory seems fascinating. Such a practice would provide as near as possible the same natural soil conditions as found in the field but would also allow more precise experimental control over the soil and plant growing conditions. Thus Biswell and Weaver (1933) and Weaver and Darland (1947) transplanted undisturbed monoliths of grass sod and underlying soil into the greenhouse to study the root development of natural grown grass sods over time.

Small undisturbed soil columns, usable for short-term root studies, can be obtained by hand core tubes. Larger soil columns may be obtained by hydraulic soil sampling machines (see Sect. 5). Details of preparing such soil columns (10 cm in diameter and up to about 140 cm in length) for use in root studies are presented by Hanus (1962) and Davis and Runge (1969). Basically, they coated the cores by dipping them in melted paraffin wax immediately after taking them from the soil.

Driving firm metal cylinders into undisturbed soil profiles and growing plants in them can generally only be recommended in sandy soils because of the difficulties in washing out the root systems from such containers (see Sect. 4.3.2). Very economical seems a technique to enwrap columns of undisturbed soil with plastic foil and to use these "constructions" for root studies (van der Post, 1968; van der Post and van der Meijs, 1968, 1969).

The technique used by soil scientists to coat the exposed surfaces of a free-standing soil monolith with a liquid plastic material which hardens and holds the monolith together (Homeyer et al., 1973) surely will be of interest also for the root ecologist in the future.

The important advantage that undisturbed soil monoliths have over refilled containers is that the existing pore space and soil texture is not disturbed and thus the soil-water relations are more like those in a natural soil profile.

10.10 Root Studies in Nutrient Solutions

Root systems grown in nutrient solutions can be observed and studied in situ without the time-consuming washing and cleaning procedure, but the growth conditions in nutrient solutions differ so extensively from those in soil that it is often difficult to compare the results obtained in soil and water (see Sect. 10.4). Roots grown in water culture are generally more fragile than those grown in soil due to their different cell structure (Knop and Wolf, 1865; Wagner, 1870).

Root studies in water culture are ideally adapted for solving physiological problems, and indeed they have been used mostly for nutrition and aeration studies. Various types of containers and apparatus with different facilities to control the environmental conditions have been described (Seiler, 1951; Martin et al., 1962; Grobbelaar, 1963; Asher et al., 1965; Geisler, 1964, 1967; De Stigter, 1969; Younis and Hatata, 1971; Kemp, 1972; Rieley and Summerfield, 1972; Kendall and Leath, 1974; Foos, 1975; Summerfield and Minchin, 1976, and others).

In recent years water cultures have been used with success for screening cultivars' response to distinct nutrients with special emphasis on root development (Fleming and Foy, 1968; Reid et al., 1971; Foy et al., 1972, and many others), and also to find out the general extent to which root systems from cultivars differ (e.g., Misra, 1956; Asana and Singh, 1967; Ruckenbauer, 1967a, 1967b, 1969).

The main advantage of growing plants in water culture is that the root system of a large number of plants can be studied simultaneously in a short time. This is one reason why several research workers in plant nutrition and plant breeding prefer water solution to soil substrate for root studies; but many replications are necessary. According to Hentrich (1966) the variability of root systems of young cereal plants grown in water culture can be very large, due to unknown external factors. He found in many experiments statistically significant differences between varieties and cultivars, but they were not always reproducable.

It would certainly be of significant benefit to plant breeders if numerous plants could be grown in water culture with a minimum of time and labour and still in the early stage of development produce reliable information on the root systems of the plants. However, in the present stage of root research it seems necessary to study the root systems of plants also in water culture for a longer period to obtain reliable results; and these results are also only of relative worth, and must be proved in soil substrate under field conditions. Today there are many data available from root studies made in water culture, but the ecological significance of adequate conclusions derived from these data cannot be evaluated, due to a lack of experimental verification under natural field conditions.

10.11 Root Studies in Mist Chambers

Growing plants under controlled conditions in mist chambers is a technique which has been used predominantly for mineral nutrition studies. It has also been used for investigating root diseases and root exudates. The mist chambers are boxes with mechanised systems for spraying nutrient solution in the boxes with an atomizer, thus establishing a continuous fog medium (Carter, 1942; Roach et al., 1957; Clayton and Lamberton, 1964; Scheffer and Kickuth, 1964; Martin and Hendrix, 1966; Przemeck and Alcalde-Blanco, 1969; Smucker and Erickson, 1976). Large root towers up to 3 m in vertical length have been constructed in the USA (Hendrix and Lloyd, 1968) and in Japan, there called "root research phytotron" (Shimura, 1970).

The development of the root systems growing in these mist chambers can be studied easily. According to Roach et al. (1957) these root systems are not covered with a slime of fungi and bacteria, as can happen in water culture. In the well-aerated mist chambers an optimal production of root hairs occurs.

Root studies in mist chambers are also more adapted for solving physiological research problems. All the data obtained in such studies must be interpreted with care, and should not be extrapolated to field conditions without adequate proof.

10.12 Comparability of Results from Container Experiments with Field Data

As emphasized by many researchers (e.g., Howard and Howard, 1918; Tornau and Stölting, 1944; Szembek, 1957, and others) and mentioned in the preceding sections, root data obtained from container-grown root systems generally should be viewed with caution, especially as related to ecological research problems. Root data obtained from water cultures or mist chambers should not be used to infer what the root distribution would be like in a natural soil profile.

Root distribution data obtained from plants grown in soil containers must be interpreted with care since in most cases the container will be too small for unrestricted root development, and root concentration at the container wall will take place. In general, it can be stated that the larger the rooting volume of the container, especially in a vertical direction, the closer the results of top and root development will be to the results obtained under natural field conditions. However, in most cases the use of ideal container size will be impractical and the data obtained can only be considered in a relative manner.

Nevertheless, root ecologists cannot do without containers for their studies since fundamental information will be obtained only by these methods. However, their main focus should center on root systems growing under natural conditions in undisturbed soil profiles and root studies in containers should only be considered as a supplement to field studies.

11. Techniques of Root Washing

11.1 Dry Sieving

Roots can be separated from soil with dry sieving only. At the field sampling site, the soil-root samples are thrown on a sloping wire screen with 2–5 mm mesh size where the roots are collected. Fresh or dry weight or other root parameters are then recorded directly without washing.

This technique has been used predominantly for studying tree roots larger then 2 mm in diameter which are not so fragile as those from herbaceous plants. Certainly, some of the finer roots are lost by this rough procedure; but the sieving is much faster than washing, and can satisfactorily solve some research problems, (Oskamp, 1933; Lyons and Krezdorn, 1962; Cockroft and Wallbrink, 1966; Samoilova, 1968).

Jutras and Tarjan (1964) developed a hydraulic soil auger-screen shaker unit for citrus root studies. The unit combines an auger for taking the soil samples and a vibrating screen, both mounted on a jeep. The soil samples are thrown on the sieve, and continuous shaking moves the soil through the screen, while the desired roots are retained on the screen surface.

The dry sieving procedure was also adapted for gaining roots of sugar cane (Lee, 1926; Lee and Bissinger, 1928) and of grasses (Roder, 1959; Bhaskaran and Chakrabarty, 1965); but in these cases the sieving was only a pre-cleaning, and the roots which were freed from the main part of the soil were later cleaned thoroughly by washing. In such pre-treated samples only root weight should be determined, not root length, due to the loss of fine roots.

Separating roots from soil by dry sieving should be applied only for sandy soils, and its use is limited to tree root studies where roots larger than 2 mm in diameter are to be investigated. In all other cases washing procedures are to be preferred.

11.2 Storing Soil-Root Samples Before Washing

It is not possible in every experiment to wash out roots from soil samples immediately after the samples have been taken from the field site. If the samples are suspended in water they can be stored about 2 to 5 days at a temperature of 15° C to 25° C before the roots start to decay. If the green parts of the plants are still connected to the roots, root growth can continue and be a source of error during storage.

If soil-root samples must be stored over a period of some weeks, ethanol or another alcohol should be added to the whole soil-root-water suspension so that the alcohol concentration is not below 10%. The amount of alcohol needed for preservation depends on the storing temperature. The higher the temperature, the higher the alcohol concentration should be. Definite recommendations for the alcohol concentrations necessary are not available. They also depend on the amount of total organic material in the suspensions.

Soil-root samples can also be preserved in dilute 4% formalin (Pavlychenko, 1942). According to Williams and Baker (1957), roots from herbaceous plants lying nine weeks in dilute formalin did not differ significantly in root weight from the original weight.

Drying is the cheapest way to preserve soil-root samples for later washing (Gliemeroth, 1952; Kmoch and Hanus, 1967; Schuurman and Goedewaagen, 1971; Opitz von Boberfeld, 1972). The samples can be air-dried or dried in an oven at about 70° C. The drying process should be not too slow because this will enhance the risk of microbial decomposition. Although the roots are shrivelled after the sample has been dried, they swell again when placed in water and partly regain their normal habit. According to Schuurman and Goedewaagen (1971) even root hairs do not seem to be harmed to a large extent.

Dried roots generally do not regain their original bright colour after they have been re-swelled in water again; they are darker than before. So in samples with large amounts of other organic material it can become more difficult to separate the roots from the debris.

An elegant method is to store the soil-root samples at temperatures below zero (Williams and Baker, 1957; Fergedal, 1967; Schuurman and Goedewaagen, 1971). The freezing absolutely stops root growth. One day before washing the soil-root samples are placed in a container with water and thawed. Freezing and thawing of the samples facilitates the washing procedure (see Sect. 4.6.4).

11.3 Chemicals for Facilitating Root Washing

It is generally not difficult to wash roots out of sandy soils; but the higher the clay content in the soil samples, the more troublesome the washing procedure will be. To facilitate the dispersion of the soil particles, several chemicals can be added to the containers in which the soil-root samples are soaked.

Generally these dispersing chemicals work most effectively when the soil samples have been partially dried before treating. They must not be totally air- or oven-dried. If soil samples rich in clay content are suspended in water and the dispersing chemical is added later to the suspension, there will be mostly little benefit.

Common chemicals for dispersing soils are pyrophosphates. Schuurman and Goedewaagen (1971) recommend the use of 0.27% sodium pyrophosphate solution (see also Sect. 4.6.4). In former times this dispersing agent was marketed under the commercial name Calgon; but today the product under this name does not contain much soluble pyrophosphate and is no longer effective for dispersing soils (Yaalon, 1976).

Sodium chloride has sometimes been used as an aid to disperse soil aggregates (Stucker and Frey, 1960; Huttel, 1975), but exact data on the salt concentrations used and their effectiveness are not outlined in the literature. McQueen (1968) soaked soil-root samples containing up to 50% organic matter in 0.5% sodium hydroxide over night. Satisfactory dispersing effects in samples which are rich in organic material or clay content have also been gained by pre-treatment with 3%–5% hydrogen peroxide solution (Dobrynin, 1968; Slavíková, 1968; Schuurman and Goedewaagen, 1971). However, pre-soaking roots for several hours in hydrogen peroxide solution can brighten dark roots, and then it becomes more difficult to distinguish new from old roots.

Effective results with clay soils containing calcium carbonate are achieved by the technique proposed by J.A. Heringa, (Wageningen, Netherlands, personal communication). He placed the unwetted soil sample in a basket made of coarse wire gauze. The basket is suspended in a box with water containing oxalic acid. This box is not much wider than the basket but is so much deeper that it allows for all the soil to pass through the bottom screen of the basket and gather underneath. The gas bubbles of carbon dioxide formed by the reaction of calcium carbonate with oxalic acid help to disperse the soil aggregates. After occasional shaking over a period of some hours, almost all the soil will fall through the bottom of the basket; the roots remain on the bottom sieve and can easily be picked out. If the soil does not contain carbonate, similar results can be reached if the samples are pre-soaked in a solution of sodium- or potassium (bi) carbonate. The most effective amount of oxalic acid depends on the kind of soil and the size of the container but in general about 10 g per litre water proved to be effective.

This technique was also used in a modified manner by Schuster and Stephenson (1940) and Carlson (1954). Steinberg and Eisenbarth (1971) and Steinberg (1973a) used 3%–5% hydrochloric acid. It seems to be a good principle to separate roots from clay soils rich in calcium carbonate without damaging the root systems.

Whether or not dispersing chemicals and cleaning agents will be a help in root washing must be decided in every individual case. When parts of the root systems are to be regained without damage to their branched roots, their use in most cases will be effective. Chemicals are of little value for cleaning tree roots which are held in place within a felty mass of mycelium (Lyford, 1975).

11.4 Washing Roots by Hand

The simplest, and in many cases still the most economic technique of separating roots from soil, is a washing process with a jet or spray of water aided by hand manipulation. In principle the soil-root sample is suspended in water and poured over fine-mesh sieves where the roots are retained and collected for further cleaning.

The round or square bottom of the sieves is made of copper gauze. The mesh size used varies from 0.2 to 2 mm^2. Which size is to be used depends on the kind of the plant root under study. If plants with very fine-branched roots are to be investigated, 0.2 mm^2 sieves should be used. Most of the research workers have used sieves with 0.5 mm^2 mesh size. For special research aims several sieves with increasing mesh size can be placed one upon another.

Very fine roots are difficult to extract and even by using a mesh size of 0.2 mm^2 roots may be lost. Caldwell and Fernandez (1975) compared two similar washing techniques, the one by suspending the root segments through a sieve with a mesh

Fig. 11.1. Washing roots on special washing tables ("Göttingen Method")

size of 0.2 mm^2, the other through 0.03 mm^2. The second technique yielded a higher biomass value by a factor of approximately 2.4. Microscopic observations indicated that this higher value was primarily due to extremely fine root tissues.

Such results cannot be generalized. Studies by Böhm (unpublished) with roots from barley plants indicated that by careful washing nearly no root segments were lost by using a sieve with a mesh size of 0.5 mm^2. Even the loss of root hairs was minimal for young plants. Such small losses occur only when plants are washed at an early stage (see Sect. 9.11). With increasing plant age roots start to decay and a small loss of root segments during the washing process occurs. This loss of roots from barley plants was in no case larger than 10% of the total root weight.

Up to now no detailed quantitative data about the loss of root segments of the various plant species during the washing process is found in the literature. It is apparent that more attention to root loss must be given by the research workers when washing out very fine and profuse root systems.

For routine root washing it is recommended to install special washing tables (Fig. 11.1). The tables should be connected with a pipe to a large pit which serves as storing room for the washed soil.

The common washing method at the Institute of Agronomy and Crop Science in Göttingen (Germany) and used in modified form also at other research places, consists of placing about 2 kg soil samples into 10-l pails. Then about 5 l water are added. After several hours, or at the next day, the pail is placed on the washing table and the soil-root-water mixture is stirred by hand until it is a homogeneous suspension. The stirring can also be done by a wooden stick, but it takes longer, and the direct control of soil particle disintegration is lost.

When the soil-root-water mixture is fully dispersed, the stirring will be interrupted for a few seconds to allow the heavy soil particles to settle down. The roots tend to float in this suspension. Then the suspension, without the settled soil particles, is poured on to the sieve. The roots remain on the sieve and the fine suspended soil particles pass through. The operation is aided by a hand sprinkler (Fig. 11.1). So the roots on the sieve are immediately freed from adhering soil suspension. Then the pail with the remaining soil is half filled again with water by means of the hand sprinkler and the process of suspension and decantation is repeated. This is done until all roots are transferred to the sieve by the decantation process. Depending on the amount of roots and soil in the sample the procedure must be repeated three to eight times.

When all roots are floated on the sieve, the remaining heavy soil particles in the pail are disposed of. In a final washing procedure the roots are transferred by means of the hand sprinkler from the sieve into the empty, clean pail, stirred, decantated again on the sieve and at last transferred with some water into the pail again ready for cleaning from debris. During the washing operation the water pressure of the hand sprinkler must not be too strong, otherwise fine or old roots can be broken into small pieces and forced through the sieve.

This "Göttingen Method" has been used by the author for many years on a large scale for root studies of agricultural plants. The time needed to wash out the roots from a 2-kg soil sample (loess soil) is about two to eight min, depending on the amount of roots in the sample. During the manual technique of stirring and suspending, roots are broken, but this does not influence the results if root weight or root length are to be determined.

For morphological investigations of root systems, e.g., if root branching is to be studied, this washing procedure is not recommended. Then it is better to place the root sample in a wire screen cylinder, and to immerse this unit in water until the soil is soft (Bloodworth et al., 1958). The cylinder is then moved up and down in the water to accelerate the separation of soil and roots. This technique is less destructive for the root system, but requires more time.

11.5 Flotation Method

With this washing technique, the soil sample is soaked in a pail of water until thoroughly dispersed. Then the entire contents of the pail is poured into a larger metal can equipped with on overflow spout (Fig. 11.2). A hose with a nozzle is placed some centimeters above the bottom of the can so that the soil-root-water mixture can circulate and overflow on to a sieve placed directly under the spout. The upward circulation of the suspension can be regulated by changing the position of the nozzle.

Fig. 11.2. Separating soil-root samples by the flotation method

The suspended soil particles which come through the overflow spout run through the sieve, while the floated roots remain on the sieve. The overflow procedure is continued by occasionally moving the hose until no more roots float out and the running water is clear.

The technique is described in detail by McKell et al. (1961) and has been used with modifications by Bommer (1955), Wetzel (1957/ 58), Long (1959), McKell et al. (1962), and Lauenroth and Whitman (1971). Without skilfully moving the hose during the overflow procedure, it is difficult to establish a continuous effective upward circulation in the overflow can, so that roots remain in the soil on the bottom of the can. It is therefore always necessary to inspect that all roots have been floated on to the sieve. These difficulties seldom occur when using a specially constructed apparatus as described by Cahoon and Morton (1961).

The floating method needs two to five times more time than the common technique of washing roots by hand, but it is less destructive for the root systems. As the separation of the roots from the soil is done under water, the root damage from strong water pressure is reduced to a minimum. Al-Khafaf et al. (1977) describe a special technique to increase the density of a dispersed soil-root suspension up to 1.5 g/cm^3 by adding pure $CaCl_2$. Roots and organic debris will float to the surface of the container and can be skimmed off with wire sieves.

The Australian method in which air-dried soil-root samples are crushed through a 5-mm sieve and soaked, and the crushed, small root segments are then collected by floating, is only to be recommended if the total macro-organic matter in a soil is to be determined (Barley, 1955; Torssell et al., 1968; Hignet, 1976). An exact separation of roots from other organic debris by this method is difficult and often impossible.

Fig. 11.3. Shaker-type root-washing machine developed by Fehrenbacher and Alexander (1955)

11.6 Root-Washing Machines

Several attempts have been made to mechanize the root-washing process. Gates (1951) proposed a technique of placing the soil-root samples into screen-bottomed cradles which were agitated in a large tank filled with water until the roots were washed out and could be collected on to a sieve. Fribourg (1953) soaked soil monoliths in screen-bottomed trays and immersed them in large drums. Subsequently he removed the trays and sprinkled them on a slatted platform from overhead nozzles.

Fehrenbacher and Alexander (1955) developed a shaker-type machine that gently put the soil-root samples into suspension by shaking and separated the roots from the soil with a sieve. This washing machine consists of a wooden stand on which a rack, containing eight pans, is free to move back and forth on rollers (Fig. 11.3). The pans, $30 \times 30 \times 45$ cm deep, are made of sheet metal with the bottoms perforated by about 1 cm holes centered 2.5 cm apart. The bottom of each pan is covered on the inside with a copper screen. The pans, when in place in the rack, extend down in a large sheet metal pan filled with water. The whole machine is driven by an electric motor. The shaking or agitation speed can be varied by using different ratios of the diameter of pulleys. A similar fluctuating washing machine was developed by Kawatake et al. (1964).

Williams and Baker (1957) constructed a washing machine with rotating sieves on which a continuous spray of water is directed to wash the soil free from the roots. Shalyt and Zhivotenko (1968) and Kolesnikov (1971) describe a Soviet root-washing machine, RWM-50, which has been designed and constructed by the Construction Bureau of the Ministry of Agriculture. Previously soaked soil samples are placed into a bunker, and moved by a conveyer on to a pan where water jets wash out a part of the soil. Then the roots with particles of soil adhering to them fall on a swinging sieve placed inside a bath where they are finally washed clean. Recently Brown and Thilenius (1976) have described a relatively low-cost root-washing machine using water spray and agitation which can be constructed from readily available commercial components for less than 400 U.S. dollars.

If man-power is the limiting factor in a root-study experiment, the use of such a root-washing machine can be effective; but often the washing procedure of a soil sample takes longer than manual operation. Comparative investigations between manual and mechanized root-washing procedures involving economy and accuracy have not been made. Thus definite recommendations for the use of the various types of washing machine cannot be given. For occasional root studies the manual washing procedure still seems the most effective.

The speed of washing a soil-root sample in such a washing machine depends on the texture, structure, and amount of organic matter of the soil. Pre-dispersion or pre-suspension of the samples speeds up the washing process considerably. The mechanized machines never wash roots cleaner than is done by manual washing. Additional cleaning with tweezers to separate debris from the roots is still necessary.

11.7 Nutrient Losses from Roots During Washing

Russell and Adams (1954) found that roots can lose phosphorus during washing, especially the finer roots which suffer the greatest mechanical injury. The authors recommend reducing these losses by washing the roots in 5%–15% solution of basic lead acetate, but such a procedure will be very costly in ecological root studies when many root samples must be washed out.

Evdokimova and Grishina (1968) also reported losses of nutrients from roots during washing. They found that roots from herbaceous plants, when subjected to prolonged washing for about 2 h had a content of calcium, manganese, and iron 10%–15% less than the total in a sample which had been washed out quickly.

According to Köhnlein and Vetter (1953), who studied root systems of various agricultural crops on a large scale, the loss of nutrients from plant roots during soaking and subsequently separating in a normal washing process can be ignored. Only from very fine roots was a small loss of potassium recorded. Also Bobritzkaja (1960) and Böhm (1973b) did not find significant losses of nutrients in the washing out of root systems of agricultural crops.

A loss of nutrients from plant roots during washing will mainly depend on how old the roots are. Brown, decaying roots will lose more nutrients than young, white roots. In ecological root studies for determination weight or length possible nutrient losses will not be a significant source of error.

The contrary effect, that too high nutrient content is found in root samples, can also occur. For example in alkaline soils calcium phosphates can be precipitated on the root surface which also remain after washing (Miller et al., 1970).

11.8 Cleaning Roots from Debris

After the roots have been separated from the soil, they must be cleaned, because in the washing procedure a separation of live roots from dead ones or from other organic matter is not possible. This cleaning is one of the most difficult tasks in ecological root research and surely the most tedious one. In early field root studies research workers often did not make this separation and still, in new publications, often no definite details about the technique of cleaning the washed root samples are given.

Mostly the aim of the cleaning procedure is to separate the live roots from dead ones and from all other organic plant debris. In routine field studies the only practical way of cleaning the washed root samples is still to sort them manually. For this purpose the washed root samples are put in flat black dishes filled with water, and the separation is done with the aid of tweezers. When there is considerable plant residue or dead roots in a sample, the separation can be facilitated by stirring the roots in the dish. Since dead plant material is lighter than living roots, it floats to the water surface quicker and can be collected with a sieve. However, this pre-separation seldom is complete. Often live roots are also floated within this debris, and must then be picked out by tweezers.

Jacques (1945) recommended using a rubber baffle as a tool for pre-separation of the roots. This baffle consists of a round flat rubber base from which cones of rubber project. By slowly pouring uncleaned root-water suspension over this baffle, the cones prevent the passage of the roots, but permit small short pieces of organic matter, dead root segments, and anything that does not cling to the rubber to be floated off. However, the roots collected on the baffle need further separation by means of tweezers, and therefore this procedure does not save time in every case.

Another possibility for separating roots from samples with huge amounts of organic debris is to boil the whole sample in water with repeated stirring. The dead organic material floats to the surface and can be picked up easily by tweezers or a sieve. However, the success of this separation depends on the degree of decomposition of the organic debris, and generally is not perfect.

Evans (1938) proposed removing plant debris and charcoal from washed samples of fine roots of sugar cane by a floating process with air bubbles in a long glass tube. The fine roots were blown up higher than the other plant debris, which was mostly deposited at the bottom of the tube when air bubbling stopped. How efficient this technique will prove for cleaning fine roots of other plant species is not known.

The main features for distinguishing live from dead roots are the root colour, the elasticity of the roots, and the presence of cortex and lateral roots (Schuurman and Goedewaagen, 1971). The combined assessments of these features can help to decide if a root should be regarded as live or dead.

In practice, however, it is often very difficult to distinguish between live and dead roots and the possibility of individual errors is great. A main emphasis is to be laid upon the root colour. The operator should be familiar with the original colour of the root system under study. Usually the roots are white, yellow, or brown, the colour depending on their age. With increasing age they change from white to brown, and when they die, they often become gray and fragile. Therefore the knowledge of the development stage of the plant under study can facilitate distinguishing live from dead roots. Also the soil temperature at which the roots have been grown can influence root colour. For example, Shanks and Laurie (1949) found that rose roots grown at 11° C were white, and those grown at 22° C were brown. Examples from other plant species are outlined by Cooper (1973).

There are rarely problems with young plants because then nearly all roots are white and easy to separate from other organic debris. Most difficulties occur if roots from perennial grasses must be cleaned. In some cases it will be nearly impossible to distinguish between live and dead roots, especially in samples from grass swards which have been taken from the surface layer of soil.

Generally the final cleaning process is done in tap water. Only if nutrient content is to be determined is it recommended to rinse finally in distilled water (Onderdonk and Ketcheson, 1973b).

To distinguish live from dead roots, several workers have used vital stains such as 2,3,5-triphenyltetrazolium chloride (Goedewaagen, 1954; Sator and Bommer, 1971; Crapo and Coleman, 1972; Knievel, 1973) and 2,3,4-triphenyltetrazolium bromide (Jacques and Schwass, 1956). Recently, Ward et al. (1978) used congo red for quantitative estimation of living wheat-root lengths in soil cores. However, as outlined in Sections 8.3 and 12.4.3 the staining techniques are involved with several problems, and until today they have been seldom used in ecological root studies in the field.

Electronically controlled devices for separating live roots from all other organic debris by photoelectric detection are still not successful today. Separating total root biomass (live and dead roots) from all other organic debris is nearly impossible when the dead roots are in the stage of decomposition. Only techniques with ^{14}C can give approximate quantitative data (see Sect. 8.5.4).

11.9 Storing Roots After Washing

If the cleaning of the roots or the determination of the root parameters in the cleaned samples cannot be done immediately after washing, the root samples must be stored. For this purpose they are placed in bottles, polyethylene bags or other plastic containers and preserved with formalin or alcohol. A 5% formalin solution or a 15%–20% alcohol solution has proved satisfactory to preserve roots for several months at temperatures of about 10° C. Meyer and Göttsche (1971) and Tennant (1976) used a mixture of alcohol, formalin, acetic acid and water.

Schuurman and Goedewaagen (1971) recommend sealing the roots with some water in plastic bags and freezing them at −20° C; but also unsealed freezing keeps roots fresh for later handling.

12. Root Parameters and Their Measurement

12.1 General Aspects

The parameters commonly used to express root growth and distribution are number, weight, surface, volume, diameter, length, and the number of root tips. The manner in which the root data is best expressed should be considered before starting an experiment. In some methods, e.g., by the use of the profile wall methods, there is little choice, but in many other study methods several approaches can be used.

Not only the research aim, but also the time and labour needed for the determination of the root parameter which has been chosen must be considered. For example, in many experiments the root ecologist would prefer to determine root surface area, but as in routine studies quick reliable determinations are not available, other parameters must be used.

There are still only very few investigations in which different root parameters have been compared in the same samples of roots. The results recently published by Böhm et al. (1977) show that the absolute values for the various root parameters can differ considerably in one experiment, but if the data are expressed as the percent distribution for various depths down the soil profile, the results obtained are similar; but these results cannot be generalised. As shown by Bloodworth et al. (1958), who compared root number and root weight, the percent root distribution down the profile can differ considerably. So for interpreting root data, it is often better to measure more than one parameter. Further research comparing various root parameters is urgently needed, and root ecologists should be encouraged to work more in this field.

12.2 Root Number

Counting root number is the common procedure when working with the core-break method (Sect.5.3), profile wall methods (Sect. 6) and glass wall methods (Sect. 7). Root numbers counted in the different soil layers generally give a good impression of the rooting density in a soil profile.

In studies where the entire undamaged root system of a single plant can be extracted the number of main and lateral roots is counted to provide an estimate of their total length (Dittmer, 1937, 1938; Pavlychenko, 1937a, 1937b); but in ecological study methods where roots are washed out from soil monoliths or soil cores, the number of roots has been counted by only few research workers, e.g., by Bloodworth et al. (1958).

Although root number is not an ideal parameter, as long and short roots are regarded and counted as equal units, high correlations with other root parameters are possible (Melhuish and Lang, 1968; Drew and Saker, 1977; Köpke, 1979).

12.3 Root Weight

12.3.1 Determination of Fresh Weight

Root weight is the most commonly used parameter for studies of root growth in response to environment. Generally the washed roots are dried and then their weight is determined. Fresh weight is frequently recorded in plant pathology in studies investigating nematodes and fungi on roots.

The simplest manner of obtaining the fresh weight is by blotting roots with blotting paper. The accuracy of this procedure is, however, influenced by individual handling, and the accuracy of the data should be controlled by subsequent measurements of the moisture content of the blotted roots. Closely reproducible fresh weights are obtained by wrapping the washed roots in muslin and additional mild centrifugation in an automatic domestic clothes washing machine (Linford and Rhoades, 1959) or by use of a low-speed bench centrifuge.

For solving other ecological problems fresh weight has very seldom been used by researchers. Thus Cockroft and Wallbrink (1966) used fresh weight for studying root systems of orchard trees. Aycock and McKee (1975) worked with fresh weight in a study of comparing several cultivars and breeding lines of tobacco.

Determination of fresh weight is simple, and by a standardized procedure for removing all adhering water from the roots, the data obtained can correlate well with other root parameters, e.g., with root dry weight or root volume (Aycock and McKee, 1975). However, on principle, if root weight is to be determined in an ecological study, dry weight should be preferred. Much information concerning the growth and function of the roots is still based on dry weight, and so with this more precise, and widely accepted parameter the results of the various research workers can be readily compared.

12.3.2 Determination of Dry Weight

To determine dry weight, the washed and cleaned roots are dried in an oven at 105° C for about 10 to 20 h depending on the amount of roots. The drying can also be done at 60° to 75° C which will take longer but can have the advantage that this lower temperature prevents roots from being pulverized (Schuurman and Goedewaagen, 1971).

Even after thorough washing and cleaning, soil particles can still adhere to the roots and cause errors. So generally it is recommended to place the dried roots into a muffle furnace at about 650° C. The cooled ash is finally treated with hydrochloric acid, filtered by decanting, dried at 105° C, and weighed.

By treating the ash residue with hydrochloric acid it is assumed that the mineral matter in the roots is dissolved. The weight of the ash residue therefore represents the weight of the unsolved soil particles in the root sample. The difference between the root dry weight, including the soil particles adhering at the roots, and the weight of the ash residue can be termed the ash-free organic root matter.

There has been dispute among research workers as to whether it is really necessary to ash roots when root dry weight is to be determined (Gericke, 1945, 1946; Bohne, 1949). Since the weight of contaminating soil particles on the washed roots can be up to 50% and more, it is advisable to determine ash-free organic matter content of the roots. Example with further details for determining root dry weight in this manner are outlined by Weiske, 1871; Willard and McClure, 1932; Köhnlein and Vetter, 1953; Könekamp, 1953; Williams and Baker, 1957; and Garwood, 1967.

The final treatment of the ash residue with hydrochloric acid is not certainly necessary in every case. As generally the root ash constitutes only a relatively small proportion of the total ash residue, its omission will be not a serious error. In soils with free lime an acid treatment should be avoided in every case because the dissolved lime is then included in the root weight (Williams and Baker, 1957). So the term ash-free organic root matter should be used in both cases whether ash residue is treated with acid or not.

Ashing of the dried root samples enhances the time and labour involved. Therefore in routine methods involving many root samples, in some cases the ashing process can be omitted. Whether this can be done depends on the kind of soil in which the plants have been grown, on the age of the roots, and on the washing technique which has been used. It is possible that in root samples the soil impurities do not exceed 4%, as found by Schuurman and Goedewaagen (1971).

12.3.3 Advantages and Critical Objections

Root weight is a good parameter for characterizing the total mass of roots in a soil. In all cases in which the productivity of underground terrestrial vegetation is to be determined, root dry weight should be the criterion for evaluation (Bray, 1963; Lieth, 1968; Bazilevich and Rodin, 1968; Santantonio et al., 1977). This includes also all research in which the contribution of the roots for the amount of humus in the soil is to be studied.

Root weight can be regarded as a fundamental measure of photosynthate storage in a plant. The known shoot-root relations are based mainly on shoot and root weight (see Sect. 12.9).

Root weight is not well adapted as a parameter for characterizing the absorbing amounts of roots in a soil. So the assumption that root weight is correlated to some kind of root activity is not valid. By expressing root data as root weight, the amount of fine roots represents only a small fraction of the total weight, although these fine roots can be the most active part of the root system. Therefore a high total root weight in a soil layer may not be identical with a zone of high water and nutrient uptake. This must be specially considered if working with plant species having thick main roots, e.g., perennial woody plants.

12.4 Root Surface

12.4.1 Calculation from Other Parameters

Surface area has seldom been determined in ecological research, although this parameter seems one of the best when experiments on water- or nutrient uptake are made.

The most direct method is to determine the average diameter of a large number of individual roots, and to measure the total root length per sample. From these data root surface can be calculated easily (Dittmer, 1937, 1938; Carlson, 1954; Kuntze and Neuhaus, 1960; Vetter and Scharafat, 1964; Schultz, 1972; Geisler and Maarufi, 1975; Adepetu and Akapa, 1977; Evans, 1977). Also by determining root diameter and root volume the surface area can be calculated (Kullmann, 1957c; Kolesnikov, 1971). But the first method is to be preferred because it needs less time.

The main drawback to these direct measurements is that the operator can decide only by inspection if the measured roots are still alive. Thus by this means an impression of the size of the root surface can be obtained, but less can be said about its activity.

12.4.2 Photoelectric Measurements

Photoelectric devices which are used, for example, for measuring leaf surface areas can also be used for estimating root surface areas. Morrison and Armson (1968) have described such a device, termed a rhizometer. In principle it consists of a light source in which the root sample is placed, a photocell for measuring the light reduction due to the roots, and a galvanometer for measuring the decrease in output from the photocell. After adjusting the device with pieces of black paper of known area, a calibration curve is made. To make measurements the roots are placed across the aperture held in one plane by a glass plate, and the readings are taken from the galvanometer. By using the calibration curve the area of root samples can be obtained. The projected area must then be converted into the total root surface by multiplying by π, assuming that the measured roots are circular in cross-section.

Working with this technique, it is important that roots do not overlap one another, and that no translucent roots are included. If translucent roots have to be measured the root sample can be stained in an opaque dye before measuring.

Several tests have shown that there exists a highly significant, linear relationship between such estimated root surface areas and the direct measurements based on measuring root diameters and root lengths. Morrison and Armson (1968) and Armson (1972) used, in studies with nursery seedlings, the data on the silhouette surface area of the roots which they termed root surface indices.

Such photoelectric devices work very well if a high proportion of fine roots do not have to be measured. If too many fine roots are present in a sample difficulties arise as Kemph (1976) has shown by using a commercial leaf area meter. The improved capabilities in direct root measurements with the QUANTIMET (see Sect. 12.7.3) encourage the hope that root samples also with large amounts of fibrous roots can be easily determined in the future.

12.4.3 Adsorption Methods

Attempts have been made to determine the root surface by dipping freshly washed roots into a dye solution and measuring the amount of dye adsorbed on the roots (Dunham, 1958; Mitchell, 1962; Kolesnikov, 1971; Kolosov, 1974). Methylene blue has been the most frequently used dye.

In principle a methylene blue solution (0.02–5 mg/l) is made and the concentration is recorded by a colorimeter. The washed root sample is then dipped into the dye solution where it remains for a standard time with gentle swirling. The adsorbed amount of methylene blue per root sample is calculated from the difference between initial and final dye concentration.

Kolosov (1974) has described a method with methylene blue solution to distinguish between active and inactive root surface. After the root sample has been immersed in the methylene blue solution it is transferred to a solution of calcium chloride, in which the methylene blue from the dead inactive root surface is displaced by calcium ions. In this way the active roots in a sample can be determined, although the data can only give approximate results about the activity of the roots for water- and nutrient uptake.

To distinguish between the total surface of a root system and its active part is an important task in root ecology, especially in tree root studies. But the known staining techniques are still beset with several problems. As found by Mitchell (1962) and by Ward et al. (1978), roots of the various plant species can react quite differently to staining. Thus, the dye adsorption is apparently related to the physical properties of the roots as well as to root cation exchange capacity. Further research is necessary to compare such root surface data obtained by staining techniques with other root parameters.

Wilde and Voigt (1949) proposed a method to determine the total root surface by titration. Washed and air-dried roots are immersed in a solution of 3 N HCl for 15 s. After the excess acid has been allowed to drain for 5 min the roots are transferred into a beaker containing 250 ml distilled water in which they remain for 10 min. An aliquot part of 10 ml of this solution is then titrated with 0.3 N NaOH, using phenolphthalein as indicator. The titration value, expressed in milliliters of NaOH, is regarded as the total capacity of the root surface.

The method can only deliver relative data about the root surface, and the same objections as outlined for the staining techniques can also be made. However, with more knowledge of cation exchange capacity of roots of various plant species, and by a careful standardization of the procedure, a practical estimation of root surface area may be possible in the future.

A simple method for determining relative root surface areas is a gravimetric method developed by Carley and Watson (1966). A beaker with a relatively viscous calcium nitrate solution is placed on a laboratory scale and weighed. Dried roots are then dipped into the calcium nitrate solution for a period of 10 s. The weight of the solution in the beaker is recorded again, and from the difference between the initial and final weight of the solution, the amount adhering to the roots can be calculated. Comparative studies made by Carley and Watson (1966) show that no significant differences in relative root surface areas exist between the titration method of Wilde and Voigt (1949) and this gravimetric method.

The results obtained with the titration method have a higher accuracy, but the gravimetric method can be suitable in routine studies due to its greater speed and simplicity. A drawback of the gravimetric method is that fine roots can rope together and behave like thicker roots (Pearson, 1974).

12.5 Root Volume

12.5.1 Calculation from Other Parameters

Root volume can be calculated by measuring the average root diameter and the root length. Such calculations, however, have seldom been done in practice (Bhaskaran and Chakrabarty, 1965). Only Vetter and Scharafat (1964) and Vetter and Früchtenicht (1969) calculated root volume per soil volume on a large scale for a large number of annual agricultural crops using the core-break method (see Sect. 5.3).

12.5.2 Displacement Technique

In general for measuring root volume the water displacement technique is to be preferred. The measuring is done in a special container with an overflow spout. This container is filled with water until it overflows from the spout. Then fresh-washed roots which have been carefully dried with a soft cloth are immersed and the overflow water volume is measured in a graduated cylinder.

Several such water-displacement devices with slight modifications have been described in the literature (Priestley and Pearsall, 1922; Novoselov, 1960; Pinkas et al., 1964; Musick et al., 1965; Andrew, 1966; Kolesnikov, 1971). Nevertheless, only a few root ecologists have used root volume as a parameter (Upchurch, 1951; Carlson, 1954; Troughton, 1963; Schmidt-Vogt, 1971; Aycock and McKee, 1975; Pickett, 1976; Atkinson et al., 1976).

The main reason for this limited use is certainly that root volume measurements from species with few large roots can be equal to species with large amounts of small fibrous roots. So the volume data alone are of limited value for most research problems in root ecology. Data from volumetric measurements in general should be used only to supplement other parameters.

12.6 Root Diameter

12.6.1 Measurements

The diameter is measured directly on freshly washed root samples with the aid of a microscope fitted with an ocular micrometer. For large roots a small hand lens, a micrometer screw or caliphers graduated to a tenth of a millimeter can be used. If individual roots differ in diameter they are measured at regular intervals throughout their length.

Before starting the measurements the roots are placed for some hours in water because many roots dry irregularly. Furthermore it must be taken into account that

Root diameter (mm)	Classes of roots	Symbol
< 0,5	Very fine	•
0.5 – 2	Fine	o
2 – 5	Small	○
5 – 10	Medium	⊙
10 – 20	Large	◎
> 20	Very large	◉

Fig. 12.1. Classes of roots with different diameters and recommended symbols for mapping them. (Modified after Melzer, 1962a)

there can be considerable diurnal variations in root diameter. On dry, sunny days the diameter of roots can shrink to about 60% of its maximum size (Huck et al., 1970).

In studies with annual plants, root diameter in most cases has been measured only for calculating root surface or root volume (see Sects. 12.4 and 12.5). But knowledge of the diameter of roots can give important information on the relationship between the pore size in a soil and the potential of root penetration (Wiersum, 1957). Examples for using root diameter as a criterion for evaluating root development under unfavourable ecological conditions are outlined by Taubenhaus and Ezekiel, 1931, Taubenhaus et al., 1931, Werner, 1931; Rauhe, 1961, 1962; Taylor et al., 1963; Aubertin and Kardos, 1965a, 1965b and Mathers, 1967. The alteration of root diameter of lettuce roots in an irrigation experiment was studied by Rowse (1974).

12.6.2 Applications in Tree Root Studies

Predominantly in tree root studies, research workers are confronted with the problem of having to differentiate between roots of various diameters. It is often convenient to divide roots into several classes according to their diameters. Published classifications differ (Evans, 1938; Karizumi, 1957, 1968; Kreutzer, 1961; Paavilainen, 1966; de Haas and Jürgensen, 1963; Moir and Bachelard, 1969; Meyer and Göttsche, 1971; Ford and Deans, 1977; Santantonio et al., 1977; further literature in Sect. 6.2.4.1). Based on the foregoing literature a division into six classes with different root diameters has been recommended in Figure 12.1.

This classification must be regarded as arbitrary, and it would be wrong to identify root diameter with any particular kind of root function. Even the assumption that the amount of very fine and fine roots, often termed fibre, fibrous, or feeder roots (Reitz and Long, 1955; Ford et al., 1957; Ford, 1963; Coker, 1959; Kaufmann et al., 1972) is closely correlated with absorbing root area is not always correct. Many roots of this category can be lignified or suberized and therefore they cannot be considered as absorbing parts (Weller, 1964, 1971).

Summarizing, the common practice of dividing tree roots into classes with different diameters is an aid to obtaining information on the amount of fine, small, medium, and large roots in a root system. However, no firm conclusions about root activity can be reached simply by sorting roots according to their diameters.

12.7 Root Length

12.7.1 Direct Measurements

For direct measurements the wet roots are placed in a flat glass dish containing a small amount of water. Graph paper, ruled in millimeters, is placed under the dish. The roots are straightened with forceps so that they do not overlap and are held in position by a glass plate. The lengths of the given roots or root segments are then estimated to the nearest millimeter by eye inspection or by the use of a magnifying glass. This kind of direct measurement has been done by Hales (1727), Nobbe (1869, 1875), Hellriegel (1883), Polle (1910), Gordienko (1930), Kolesnikov (1930, 1962a, 1971), Kokkonen (1931), Dittmer (1937, 1938, 1948, 1959b), Pavlychenko (1937a, 1937b), Grosskopf (1950), Spencer (1951), Kern et al. (1961), Strydom (1964), and Cockroft and Wallbrink (1966).

Instead of arranging the roots randomly in a dish every individual root can be dipped into gum, and mounted end to end in lines on graph paper ruled in millimeters. Branched roots are cut into individual segments before glueing, and handled in the same manner as described above (Nutman, 1934; Thomas, 1944; Stankov, 1960; Barua and Dutta, 1961).

Obtaining root-length data by these direct measurements is tedious and time-consuming, and the procedure is recommended only for estimating the length of single roots, not as a routine method in ecological research.

12.7.2 Intersection Methods

Instead of tedious direct measurements, root length can be calculated more rapidly by counting the intersections between roots and a regular pattern of lines. Head (1966) has used this kind of line intersection method in the East Malling Root Laboratory (England). He counted the total number of intersections between the roots and the vertical and horizontal lines of a grid on the glass observation windows. Comparisons of estimated intersection data with the measured actual root length show a linear relation between the number of intersections and the actual root length (Fig. 12.2).

Independently of this practical line intersection method, Newman (1966a) developed a theory that root length can be estimated by the equation

$$R = \pi AN/2H$$

where R is the total length of roots in a field of area A and N is the number of intersections between the roots and random straight lines of total length H. In practice the roots are laid out on a flat surface, and the number of intersections

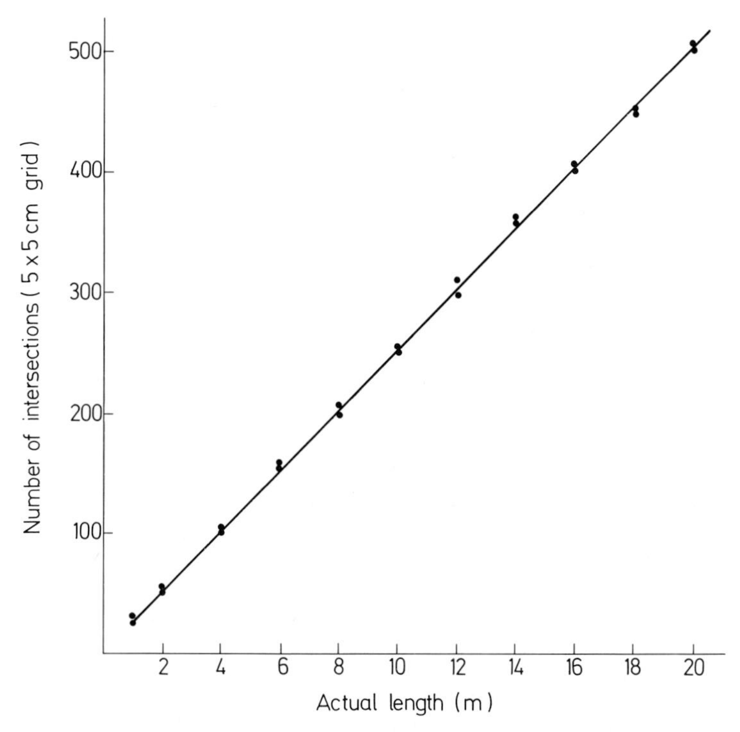

Fig. 12.2. The relationship between actual and estimated length of cotton thread recorded by the direct method and by the intersection method (r = 0.99)

between the roots and the random straight lines is counted. In the original Newman technique a microscope hair-line provides the straight lines.

In recent years Newman's method has been modified and improved by several research workers (Torssell et al., 1968; Evans, 1970; Marsh, 1971; Tennant, 1975). The main change is that for the area over which the roots are spread any convenient size of grid system can be used.

Today root length can be determined without any expensive equipment in a shallow dish made of transparent plastic or glass. A convenient dish size is 30 × 40 cm. A grid is placed under the bottom of the dish. The wet roots are then poured into the dish with some water, and they are positioned randomly over the grid with forceps or needles so that they do not overlap (Fig. 12.3). If necessary, long branched roots are cut into smaller pieces. Finally, counts are made of the intersections of the roots with the vertical and horizontal grid lines. The use of a hand tally counter facilitates the counting procedure.

What size of grid should be used depends on the amount of roots to be measured. For small root samples with total length below 1 m a 1-cm grid, for larger samples to about 5 m a 2-cm grid, and for total root lengths up to 15 m a 5-cm grid is recommended. After the experiences of Köpke (1979) the number of intersections to be counted in one root sample should not exceed 400, due to possible fatigue of the operator. But the counted number of intersections should also not be below 50 because then the accuracy of the results decreases.

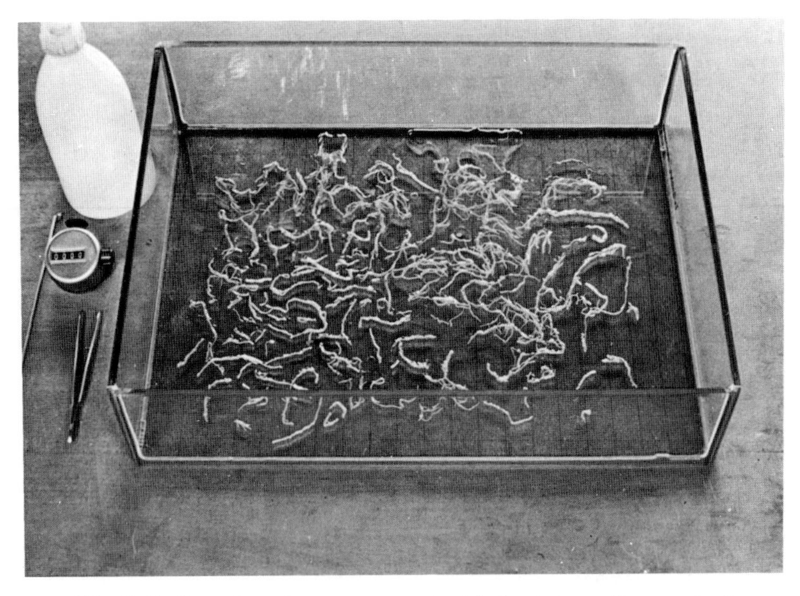

Fig. 12.3. Root length measurement with the intersection method

Based on the consideration of Marsh (1971) recently Tennant (1975) argued that Newman's (1966) formula can be simplified. For a grid of indeterminate dimensions the intersection counts can be converted to centimetre measurements using the equation:

$$\text{Root length (R)} = \tfrac{11}{14} \times \text{Number of intersections (N)} \times \text{Grid unit}$$

As proposed by Tennant (1975) the $\frac{11}{14}$ of the equation can be combined with the grid unit, and thus a length conversion factor is obtained. The factors for the 1-, 2-, and 5-cm grid squares are 0.786, 1.57, and 3.93 respectively. These conversion factors can also be derived from the diagrams first proposed by Head (1966) which show the linear relationship between actual root length and the number of intersections (see Fig. 12.2). In practice there are now no differences between the modified Newman method recently described by Tennant (1975), and the intersection method proposed by Head (1966).

Practical tests done by Head (1966), Newman (1966a), Reicosky et al. (1970), Tennant (1975), and others indicate that there are only small differences in the precision between the data from the direct measurements and those obtained from the intersection method. In no case has the standard deviation been more than $\pm 10\%$ of the true lengths, and in most cases it has been much less.

A source of error can be the counting of curved root segments. Tennant (1975) obtained good results by recording single counts when the edges of the curved roots touched a line, and recording double counts in cases where roots were lying along a grid line.

12.7.3 Root-Counting Machines

An instrument which utilizes the line intersection principle was developed by Rowse and Phillips (1974). The intersections between the roots and parallel lines are counted by moving an arranged root sample on a glass plate beneath a special binocular microscope fitted with a photo-electric counting device. Each time a root passes beneath the microscope, a count is registered electronically by a scaler. The actual root length can be read directly on the scaler.

Root length measurements with this instrument are done more rapidly than by manual counting with a hand tally counter. Furthermore, as the instrument works automatically, the operator can arrange the next root sample while the counting is being done by the machine. The limiting factor usually is the time required to separate and arrange the roots on the glass plate. This must still be done by hand.

It is expected that root length measurements will be facilitated in the future by using the image analysing computer QUANTIMET 720 manufactured by Cambridge Instruments, Melbourn, Royston, Herts., England. This electro-optical image analysis equipment determines space and shape parameters in any given pattern by a scanning technique (Fisher, 1971).

For practical determinations, the roots are arranged in a flat dish and photographed against a strongly contrasting background. From the 35 mm slides obtained the computer measures the total root length in seconds. First experiments with the QUANTIMET 720 have been reported by Baldwin et al. (1971). The time required to prepare high-quality photographs is still a limitation in routine studies.

12.7.4 Reasons for Increasing Use of Root Length Measurements

During the last decade an increasing number of research workers have been using length as the preferred measure in root studies. One reason for this tendency is the new possibilities for rapid determination. More importantly, research workers believe that root length per unit soil volume is one of the best parameters for calculations of water uptake by plant roots (Grosskopf, 1950; Gardner, 1964; Molz, 1971; Taylor and Klepper, 1973, 1975).

Also for studying the process of nutrient uptake by plant roots the length of the roots seems to be a good parameter, as indicated in a theoretical work by Nye and Tinker (1969). Many research workers in the field of plant nutrition now prefer root length to root weight or other parameters (e.g., Khasawneh and Copeland, 1973; Soileau, 1973; Claassen and Barber, 1974; Mengel and Barber, 1974b; Fitter, 1976). However, also in the broad realm of various ecological field studies in recent years more research workers have been used root length as a parameter (Reynolds, 1970, 1974, 1975; Barber, 1971; Schultz, 1974; Welbank et al., 1974; Grimes et al., 1975, 1978; Böhm, 1976; Goodman and Greenwood, 1976; Hignet, 1976; Lesczynsky and Tanner, 1976; Scott and Oliver, 1976; Ambler and Young, 1977; Böhm et al., 1977; Sivakumar et al., 1977; Gregory et al., 1978; Köpke, 1979).

12.8 Root Tips

12.8.1 Technique of Counting

For determination, the cleaned roots are placed in a flat dish where they are kept under water. Here they are arranged so that they do not overlap, as described in Section 12.7. Then the number of live root tips is counted using a stereomicroscope.

The distinction between live and dead root tips can be made by evaluating their morphology and colour. Root tips are generally assumed to be alive if they are turgid, and white to light brown in colour (Weller, 1971).

Although from the ecological standpoint root tips seem to be a relatively good parameter (surely better than root weight) only a few research workers have used them, predominantly in tree root studies (Morrow,1950; Kalela, 1955; Weller, 1964, 1966a, 1966b, 1967, 1968, 1971; Meyer, 1967; Steinberg and Eisenbarth, 1971; Meyer and Göttsche, 1971; Steinberg, 1972, 1973a, 1973b; Fritzsche and Nyfeler, 1974; Vávra, 1975).

The main limitation seems to be that the counting procedure is very time-consuming and therefore not well adapted for routine studies. It is advisable for every research worker who intends to count root tips to limit the samples to the lowest possible number that will still be acceptable for statistical calculations (Vávra, 1975).

12.8.2 Root Coefficients

To characterize the root branching of woody plants, some research workers have proposed the use of simple coefficients, in which the number of root tips are compared with other root parameters.

Schreiber (1926) and Hesselink (1926) used the coefficient (R):

$$R = \frac{\text{mm root length}}{\text{number of root tips}}$$

This coefficient was later modified by Stellwag-Carion (1937), and by Otto (1962, 1963, 1964).

Frischenschlager (1935) proposed the coefficient (K):

$$K = \frac{\text{g root weight}}{\text{number of root tips}}$$

Steinberg and Eisenbarth (1971) and Steinberg (1973a, 1973b) also used root weight, but their coefficient was the inverse.

Although these coefficients can give valuable information on the branching density of a root system, they are not widely used in root ecology today. Nevertheless, the branching density of root systems and its variation under different ecological conditions should be a field of further research, and it is hoped that some research workers will not be discouraged by the time- and labour-consuming measurements which are necessary.

12.9 Shoot-Root Relations

The common parameter for evaluating the relations between above- and belowground growth of plants is the shoot:root ratio. It is a measure of the distribution of dry weight between shoot and root system of the plants. Boonstra (1931, 1955) proposed using the coefficient (C) obtained by the equation:

$$C = \frac{\text{shoot weight}}{\text{root weight}}$$

Other research workers (e.g., Bray, 1963; Wilson, 1975) used the inverse coefficient.

Since generally the dry weight of the aboveground parts of plants is larger than that of the roots, in the first case the value will be >1, in the latter <1. Könekamp (1936, 1953) and Bommer (1955) who also used the inverse coefficient, multiplied the results by 100, to give values >1. Most research workers seem to prefer the equation proposed by Boonstra.

Selected studies on the shoot–root relations of grasses and herbaceous plants are those by Hosäus (1872), Opitz (1904), Livingston (1906), Schulze (1906, 1911/14), Schneider (1912), Harris (1914), Clausen (1929), and Derick and Hamilton (1942). In recent years fundamental basic research have been done by Troughton (1955, 1956, 1960, 1963, 1968, 1974, 1977), Brouwer et al. (1961), Brouwer (1963, 1967), Steineck (1964), Ruckenbauer (1967a, 1967b, 1969), Hunt (1975, 1976), and Hunt et al. (1975). Details and problems of shoot–root relations of trees are outlined by Hilkenbäumer (1959), Barlow (1960), Melzer (1962b), Kira and Ogawa (1968), Ledig et al. (1970), and Wilson (1975).

Many of the results of all this research indicate that the common assumption of a persistant tendency towards a positive correlation between shoots and roots is not generally valid. Ecological conditions can change the shoot:root ratio considerably when plants are growing under field conditions (Köhnlein and Vetter, 1953; Gäde, 1962). In the present stage of research the statement of Roberts and Struckmeyer (1946) still seems to be valid that the shoot:root ratio on a dry weight basis will offer less than has sometimes been expected towards a solution of the efficiency of the roots for plant growth.

Nevertheless, studying shoot–root relations should be a central research field for root ecologists. Adequate information is essential for analyzing or simulating growth pattern of whole plants (Mayaki et al., 1976; Sivakumar et al., 1977) and for estimating primary productivity of ecosystems (Lieth and Whittacker, 1975). However, when considering the efficiency of a root system for plant growth, root distribution in the soil profile, rather than total root weight per se seems the more important factor. This should be taken into account in future research. A step in this direction has been made recently by Parao et al. (1976). By testing the drought resistance of upland rice varieties, which depends on a well proliferated deep root system, they obtained significant correlations between the growth of shoots and of roots growing in the soil layers below 30 cm.

Studies of shoot–root relations in the future should not be restricted to measurement of dry weight alone. The increasing use of length as a root parameter encourages the possibility that relationships will be found between root length and development stages of the plant.

In addition to all studies on ecological factors which seem to control the shoot–root relations, research workers should keep up-to-date with the results from physiological research on the hormonal mechanisms controlling the growth of plants. The new findings by root physiologists indicate a significant hormonal feedback control between shoot and root system (Whittington, 1969; Sytnik, 1972; Carson, 1974; Hoffmann, 1974a; Kolek, 1974; Torrey and Clarkson, 1975; Russell, 1977). In the field of research on shoot–root relations root ecologists and root physiologists will probably have the closest contact in the future.

13. Some Future Aspects for the Use of Ecological Root-Study Methods

Knowledge of the plant root systems is the key to fundamental ecological understanding in many fields of botany and their applied sciences. Still relatively less is known about roots than about aboveground parts of the plants, as often stated in summarizing research reviews (Brouwer, 1966; Rogers and Head, 1966; Russell, 1971). In the past most of the information regarding root growth in the field has been obtained only incidentally from studies with other objectives (Wiersma, 1959). However, predominantly since the last decade, there has been a significant tendency to fill out the gaps in knowledge by better-directed root research.

In root ecology the main research field in the future will continue to obtain more quantitative data about the natural ecological conditions that are favourable or unfavourable for the root systems, for example their reaction to soil compaction, soil aeration, or moisture status. For cultivated crops it is important to obtain more detailed information on the factors which impede root growth in the subsoil, and thus to decrease the risk of lower yields. Further studies to find pioneer plants which can relatively easily gain access to subsoil layers can be of practical interest in agronomy (Köhnlein, 1955, 1960; Köhnlein and Bergt, 1971).

A wide open field is the study of the root distribution in and between the rows of cultivated row crops to give hints for improving fertilizing and irrigation techniques. Examples of the various applications of ecological root research in agronomy are summarized by Pätzold (1963a, 1963b), in horticulture by Crider (1927), Weller (1964), and Kolesnikov (1962b, 1969, 1971), and in forestry by Laitakari (1929b), Brückner (1950), Wagenknecht (1955), and Köstler et al. (1968).

In connection with determinations of the total biomass production and their development in ecosystems (Lieth and Whittaker, 1975; Marshall, 1977) more data on the root mass will lead to better knowledge of shoot–root relations of the plants.

Plants with a deep-growing root system will generally survive drought periods better than those with a shallow root system. In regions where water is a limiting factor, rooting depth is an important factor for satisfactory crop production (Rotmistrov, 1926; Ostermayer, 1934; Parsche, 1941; Geisler, 1957; Pearson, 1974). Studies to find differences in the drought resistance governed by different root systems between crop cultivars have been done on a large scale in India (Bose and Dixit, 1931; Subbiah et al., 1968; Thangavelu et al., 1969; Katyal and Subbiah, 1971; Bhan et al., 1973) and in Canada (Hurd, 1964, 1968, 1974; Hurd and Spratt,

1975). They have been started also at many other places in the world (e.g., Derera et al., 1969; Lupton et al., 1974; Parao et al., 1976; Taylor et al., 1978). In the present stage of methodological possibilities in root ecology, this kind of research will in most cases be a question of selecting among the existing well-known cultivars (Böhm, 1973a). The value to the plant breeder of the genetic variation of the root systems in breeding programmes is still controversial as quick routine methods are lacking (Troughton and Whittington, 1969; Monyo and Whittington, 1970; Zobel, 1975).

In the future there will be an increasing demand for root data from workers developing mathematical models (e.g., Brouwer and De Wit, 1969; Lungley, 1973; Ares and Singh, 1974; Gerwitz and Page, 1974; Taylor and Klepper, 1975; Hillel et al., 1976). It is hoped that the dialogue between these scientists and the root ecologists doing practical work in the field will hasten the process of understanding the complex mechanism of shoot–root relations. Conclusions made by Innis (1977) suggest that modelling may play a more active role in the study of the root systems than it has in the shoot systems.

As a general rule, it will be necessary in the future to engage in ecological root research predominantly on a dynamic basis. Data from root studies where the root system has been measured only once will be of only limited value. Therefore the use of glass wall methods will certainly increase.

Most of the future ecological root research will, however, still be done with the direct traditional field methods. Increasing costs of employing assistants will certainly accelerate the process of mechanized sampling, washing, and measuring the roots; but the principles of most of the methods described will probably be the same for the next few decades. Revolutionary new root-study methods are not in sight at the present time.

References

Aaltonen, V.T.: Über die Ausbreitung und den Reichtum der Baumwurzeln in den Heidewäldern Lapplands. Acta For. Fenn. *14*, 1–55 (1920)

Adepetu, J.A., Akapa, L.K.: Root growth and nutrient uptake characteristics of some cowpea varieties. Agron. J. *69*, 940–943 (1977)

Adriance, G.W., Hampton, H.E.: Root distribution in citrus, as influenced by environment. Proc. Am. Soc. Hortic. Sci. *53*, 103–108 (1949)

Agrawal, R.P., Khanna, R.K., Nath, J., Batra, M.L.: Root penetration studies with ^{32}P in cereals as affected by compact layers at varying depths. Ann. Arid Zone *14*, 339–346 (1975)

Ahenkorah, Y.: A study of the distribution of root activity of mature cacao (*Theobroma cacao* L.) using the P 32 soil injection technique. Ghana J. Agric. Sci. *2*, 97–101 (1969)

Albrecht, D.: Verbesserung der Spatendiagnose. Dtsch. Landwirtsch. *2*, 41–43 (1951)

Albrecht, D., Fritzsche, K.H., Winkler, S.: Weitere Entwicklung des Strukturbohrers. Dtsch. Landwirtsch. *4*, 206–208 (1953)

Aldrich, W.W., Work, R.A., Lewis, M.R.: Pear root concentration in relation to soil moisture extraction in heavy clay soil. J. Agric. Res. *50*, 975–988 (1935)

Al-Khafaf, S., Wierenga, P.J., Williams, B.C.: A flotation method for determining root mass in soil. Agron. J. *69*, 1025–1026 (1977)

Allmaras, R.R., Nelson, W.W.: Corn (*Zea mays.* L.) root configuration as influenced by some row-interrow variants of tillage and straw mulch management. Proc. Soil Sci. Soc. Am. *35*, 974–980 (1971)

Allmaras, R.R., Nelson, W.W., Voorhees, W.B.: Soybean and corn rooting in Southwestern Minnesota: II. Root distributions and related water inflow. Proc. Soil Sci. Soc. Am. *39*, 771–777 (1975)

Ambler, J.R., Young, J.L.: Techniques for determining root length infected by vesicular-arbuscular mycorrhizae. J. Soil Sci. Soc. Am. *41*, 551–556 (1977)

Andrew, R.H.: A technique for measuring root volume in vivo. Crop Sci. *6*, 384–386 (1966)

Angelo, E., Potter, G.F.: The error of sampling in studying distribution of the root systems of tung trees by means of the Veihmeyer soil tube. Proc. Am. Soc. Hortic. Sci. *37*, 518–520 (1939)

Ares, J., Singh, J.S.: A model of the root biomass dynamics of a shortgrass prairie dominated by blue grama (*Bouteloua gracilis*). J. Appl. Ecol. *11*, 727–744 (1974)

Arkin, G.F., Blum, A., Burnett, E.: A root observation chamber field installation. Texas Agricultural Experiment Station, Texas, USA. Misc. Publ. No. 1386. 1978

Armson, K.A.: Distribution of conifer seedling roots in a nursery soil. For. Chron. *48*, 141–143 (1972)

Asana, R.D., Singh, D.N.: On the relation between flowering time, root growth and soil-moisture extraction in wheat under non-irrigated cultivation. Indian J. Plant Physiol. *10*, 154–169 (1967)

Asher, C.J., Ozanne, P.G.: Root growth in seedlings of annual pasture species. Plant Soil *24*, 423–436 (1966)

Asher, C.J., Ozanne, P.G., Loneragan, J.F.: A method for controlling the ionic environment of plant roots. Soil Sci. *100*, 149–156 (1965)

Atkinson, D.: Seasonal periodicity of black currant root growth and the influence of simulated mechanical harvesting. J. Hortic. Sci. *47*, 165–172 (1972)

Atkinson, D.: Seasonal changes in the length of white unsuberized root on raspberry plants grown under irrigated conditions. J. Hortic. Sci. *48*, 413–419 (1973)
Atkinson, D.: Some observations on the distribution of root activity in apple trees. Plant Soil *40*, 333–342 (1974)
Atkinson, D.: Some observations on the root growth of young apple trees and their uptake of nutrients when grown in herbicided strips in grassed orchards. Plant Soil *46*, 459–471 (1977)
Atkinson, D., Naylor, D., Coldrick, G.A.: The effect of tree spacing on the apple root system. Hortic. Res. *16*, 89–105 (1976)
Aubertin, G.M.: Nature and extent of macropores in forest soils and their influence on subsurface water movement. U.S.D.A. For. Serv. Res. Pap. NE-192. Northeast. For. Exp. St. Upper Darby, PA, USA (1971) 33 pp
Aubertin, G.M., Kardos, L.T.: A method of transplanting and supporting seedlings for short time research studies. Agron. J. *56*, 523 (1964)
Aubertin, G.M., Kardos, L.T.: Root growth through porous media under controlled conditions. I. Effect of pore size and rigidity. Proc. Soil Sci. Soc. Am. *29*, 290–293 (1965a)
Aubertin, G.M., Kardos, L.T.: Root growth through porous media under controlled conditions. II. Effect of aeration levels and rigidity. Proc. Soil Sci. Soc. Am. *29*, 363–365 (1965b)
Aycock, M.K., McKee, C.G.: Root size variability among several cultivars and breeding lines of Maryland tobacco. Agron. J. *67*, 604–606 (1975)
Bakermans, W.A.P., De Wit, C.T.: Crop husbandry on naturally compacted soils. Neth. J. Agric. Sci. *18*, 225–246 (1970)
Balázs, F.: Investigations on the root development of cereals. (Russ., Engl.Sum.). Acta Agron. Acad. Sci. Hung. *4*, 69–103 (1954)
Baldwin, J.P., Tinker, P.B.: A method for estimating the lengths and spatial patterns of two interpenetrating root systems. Plant Soil *37*, 209–213 (1972)
Baldwin, J.P., Tinker, P.B., Mariott, F.H.C.: The measurement of length and distribution of onion roots in the field and the laboratory. J. Appl. Ecol. *8*, 543–554 (1971)
Ballantyne, A.B.: Fruit Tree Root Systems. Spread and Depth. Utah Agric. Coll. Exp. Stn. Bull. No. 143 (1916) 15 pp
Bär, K., Tseretheli, O.: Der Einfluss der Schnitthäufigkeit auf die Wurzelentwicklung junger Luzerne. Pflanzenbau *19*, 317–328 (1943)
Barber, S.A.: Effect of tillage practice on corn (*Zea mays.* L.). Root distribution and morphology. Agron. J. *63*, 724–726 (1971)
Barber, S.A.: Growth and nutrient uptake of soybean roots under field conditions. Agron. J. *70*, 457–461 (1978)
Bargioni, G.: A study of cherry root systems in the Verona district. (Ital., Engl. Sum.). Riv. Ortoflorofruttic. Ital. *43*, 100–119 (1959a)
Bargioni, G.: Studies and research on the root systems of peach trees in the province of Verona. (Ital., Engl. Sum.). Riv. Ortoflorofruttic. Ital. *43*, 400–419 (1959b)
Barley, K.P.: The determination of macro-organic matter in soils. Agron. J. *47*, 145–147 (1955)
Barley, K.P., Farrell, D.A., Greacen, E.L.: The influence of soil strength on the penetration of loam by plant roots. Aust. J. Soil Res. *3*, 69–79 (1965)
Barlow, H.W.B.: Root-shoot relationships in fruit trees. Sci. Hortic. *14*, 35–41 (1960)
Bartos, D.L., Sims, P.L.: Root dynamics of a shortgrass ecosystem. J. Range Manage. *27*, 33–36 (1974)
Barua, D.N., Dutta, K.N.: Root growth of cultivated tea in the presence of shade trees and nitrogenous manure. Emp. J. Exp. Agric. *29*, 287–298 (1961)
Bassett, D.M., Stockton, J.R., Dickens, W.L.: Root growth of cotton as measured by P^{32} uptake. Agron. J. *62*, 200–203 (1970)
Batchelder, A.R., Bouldin, D.R.: Technique for determining root elongation rates through soil layers of different physical and chemical properties. Agron. J. *64*, 49–52 (1972)
Bates, G.H.: A device for the observation of root growth in the soil. Nature (London) *139*, 966–967 (1937)

Batjer, L.P., Oskamp, J.: Soils in relation to fruit growing in New York. Part VII. Tree behavior on important soil profiles in the Kinderhook, Germantown, and Red Hook areas in Columbia and Dutchess Counties. Cornell Univ. Agric. Exp. Stn. Ithaca, New York. Bull. 627 (1935) 30 pp

Baumann, H.: Wasserversorgung und Wurzelbildung. Dtsch. Landwirtsch. *2*, 65–67 (1948)

Baumann, H., Klauss, M.L.: Über die Wurzelbildung bei hohem Grundwasserstand. Z. Acker- Pflanzenbau *99*, 410–426 (1955)

Bausch, W., Onken, A.B., Wendt, C.W., Wilke, O.C.: A self-propelled high-clearance soil coring machine. Agron. J. *69*, 122–124 (1977)

Bazilevich, N.I., Rodin, L.E.: Reserves of organic matter in underground sphere of terrestrial phytocoenoses. In: Methods of Productivity Studies in Root Systems and Rhizosphere Organisms. Int. Symp. USSR 1968. Ed. by USSR Academy of Sciences, Leningrad: Nauka, 1968, pp. 4–8

Beard, F.H.: Root Studies X. The root-systems of hops on different soil types. J. Pomol. Hortic. Sci. *20*, 147–154 (1943)

Becking, J.H.: Root nodules in non-legumes. In: The Development and Function of Roots. Torrey, J.G., Clarkson, D.T. (eds.) London: Academic Press, 1975, pp. 507–566

Bennett, O.L., Doss, B.D.: Effect of soil moisture level on root distribution of cool-season forage species. Agron. J. *52*, 204–207 (1960)

Bergmann, W.: Wurzelwachstum und Ernteertrag. Z. Acker-Pflanzenbau *97*, 337–368 (1954)

Bergmann, W.: Über die Beeinflussung der Wurzelbehaarung von Roggenkeimpflanzen durch verschiedene Außenfaktoren. Z. Pflanzenernähr. Düng. Bodenk. *80*, 218–224 (1958)

Bhan, S., Singh, H.G., Singh, A.: Note on root development as an index of drought resistance in sorghum [*Sorghum bicolor* (L.) Moench.]. Indian J. Agric Sci. *43*, 828–830 (1973)

Bhar, D.S., Mason, G.F., Hilton, R.J.: Microscope trolley for rhizotron. Can. J. Plant Sci. *49*, 104–106 (1969)

Bhar, D.S., Mason, G.F., Hilton, R.J.: In situ observations on plum root growth. J. Am. Soc. Hortic. Sci. *95*, 237–239 (1970)

Bhaskaran, A.R., Chakrabarty, D.C.: A preliminary study on the variations in the soil binding capacity of some grass roots. Indian J. Agron. *10*, 326–330 (1965)

Bilan, M.V.: Root development of loblolly pine seedlings in modified environments. Dep. For. Stephen F. Austin State Coll. Nacogdoches, Texas, USA. Bulletin No. 4 (1960) 31 pp.

Bilan, M.V.: Acrylic resin tubes for studying root growth in tree seedlings. For. Sci. *10*, 461–462 (1964)

Billings, W.D., Shaver, G.R., Trent, A.W.: Measurement of root growth in simulated and natural temperature gradients over permafrost. Arct. Alp. Res. *8*, 247–250 (1976)

Biswell. H., Weaver, J.E.: Effect of frequent clipping on the development of roots and tops of grasses in prairie sod. Ecology *14*, 368–389 (1933)

Blaauw, A.H.: Small building-constructions for physiological cultivation-experiments. (Dutch, Engl. Sum.). Meded. Landbouwhogesch. Wageningen *25* (3), 1–20 (1923)

Blaser, R.E.: A rapid quantitative method of studying roots growing under field conditions. J. Am. Soc. Agron. *29*, 421–423 (1937)

Bloodworth, M.E., Burleson, C.A., Cowley, W.R.: Root distribution of some irrigated crops using undisrupted soil cores. Agron. J. *50*, 317–320 (1958)

Bloomberg, W.J.: Two techniques for examining root distribution. Can. J. Plant Sci. *54*, 865–868 (1974)

Blydenstein, J.: Root systems of four desert grassland species on grazed and protected sites. J. Range Manage. *16*, 93–95 (1966)

Boatwright, G.O., Ferguson, H.: Influence of primary and/or adventitious root systems on wheat production and nutrient uptake. Agron. J. *59*, 299–302 (1967)

Bobritzkaja, M.A.: Einfluß der Tiefenbearbeitung auf die biologischen Eigenschaften des Bodens und die Entwicklung des Wurzelsystems sowie damit zusammenhängende Fragen der Untersuchungsmethodik. Dtsch. Akad. Landwirtschaftswiss. Berlin. Tagungsber. *28*, 195–198 (1960)

Bockemühl, J.: Entwicklungsweisen des Klatschmohns im Jahreslauf als Hilfen zum Verständnis verwandter Arten. Elem. Naturwiss. *19*, 37–52 (1973)
Bodo, F.: Untersuchungen auf dem Gebiete des Wurzelwachstums des Apfels und der Zwetschke. Fortschr. Landwirtsch. *1*, 768–773 (1926)
Boehle, J., Mitchell, W.H., Kresge, C.B., Kardos, L.T.: Apparatus for taking soil-root cores. Agron. J. *55*, 208–209 (1963)
Boeker, P.: Die Wurzelentwicklung unter Rasengräserarten und -sorten. Rasen-Turf-Gazon *5*, 1–3, 44–47, 100–105 (1974a)
Boeker, P.: Die Wurzelmassenentwicklung einiger Untergräser. Wirtschaftseigene Futter *20*, 82–94 (1974b)
Böhm, W.: Wurzelforschung mit Polyäthylen-Röhren. Plant Soil *37*, 683–687 (1972)
Böhm, W.: Wurzelwachstum und Ertragsbildung bei Körnerfrüchten. Vortr. Pflanzenzüchter *13*, 152–164 (1973a)
Böhm, W.: Wurzelwachstum bei Sommergerste in Abhängigkeit von der P-Düngerplacierung in phosphatarmen Lössböden. Z. Acker-Pflanzenbau *138*, 99–115 (1973b)
Böhm, W.: Phosphatdüngung und Wurzelwachstum. Phosphorsäure *30*, 141–157 (1974a)
Böhm, W.: Mini-rhizotrons for root observations under field conditions. Z. Acker-Pflanzenbau *140*, 282–287 (1974b)
Böhm, W.: Wurzelforschung und Landschaftsökologie. Nat. Land. *49*, 158–161 (1974c)
Böhm, W.: In situ estimation of root length at natural soil profiles. J. Agric. Sci. *87*, 365–368 (1976)
Böhm, W.: Development of soybean root systems as affected by plant spacing. Z. Acker- Pflanzenbau *144*, 103–112 (1977)
Böhm, W.: Die Bestimmung des Wurzelsystems am natürlichen Standort. Kali-Briefe *14*, 91–101 (1978a)
Böhm, W.: Untersuchungen zur Wurzelentwicklung bei Winterweizen. Z. Acker-Pflanzenbau *147*, 264–269 (1978b)
Böhm, W., Köpke, U.: Comparative root investigations with two profile wall methods. Z. Acker- Pflanzenbau *144*, 297–303 (1977)
Böhm, W., Maduakor, H., Taylor, H.M.: Comparison of five methods for characterizing soybean rooting density and development. Agron. J. *69*, 415–419 (1977)
Böhme, H.: Untersuchungen über die Bewurzelung der Industriekartoffel. J. Landwirtsch. *73*, 81–144 (1925)
Böhme, H.: Die Bedeutung der Wurmröhren für das Tiefenwachstum der Wurzeln. Pflanzenbau *3*, 139–143 (1926/27)
Böhme, H.: Beiträge zur Wurzelforschung. Pflanzenbau *4*, 56–62, 72–78 (1927/28)
Bohne, H.: Bemerkungen zu den Arbeiten von S. Gericke über das Thema "Die Bedeutung der Ernterückstände für den Humushaushalt des Bodens". Z. Pflanzenernähr. Düng. Bodenkde. *44*, 65–71 (1949)
Bohne, H., Garvert, J.: Untersuchungen über die Bedeutung der Ernterückstände des Getreides für die Humusversorgung. Z. Pflanzenernähr. Düng. Bodenkde. *55*, 170–178 (1951)
Bohne, H., Greiffenberg, H.: Mineraldüngung und Wurzelwachstum bei S.-Gerste und Hafer. Z. Pflanzenernähr. Düng. Bodenkde. *64*, 67–72 (1954)
Bojappa, K.M., Singh, R.N.: Root activity of mango by radiotracer technique using ^{32}P. Indian J. Agric. Sci. *44*, 175–180 (1974)
Bole, J.B.: Influence of root hairs in supplying soil phosphorus to wheat. Can. J. Soil Sci. *53*, 169–175 (1973)
Bommer, D.: Untersuchungen über die Ernterückstände von Feldfutterpflanzen in verschiedenen Höhenlagen. Z. Acker- Pflanzenbau *99*, 239–258 (1955)
Boogie, R., Knight, A.H.: Studies of root development in a grass sward growing on deep peat using radioactive tracers. J. Br. Grassl. Soc. *15*, 133–136 (1960)
Boogie, R., Hunter, R.F., Knight, A.H.: Studies of the root development of plants in the field using radioactive tracers. J. Ecol. *46*, 621–639 (1958)
Boonstra, A.E.: Pflanzenzüchtung und Pflanzenphysiologie. Züchter *3*, 345–352 (1931)

Boonstra, A.E.: Het begrip "wortelwaarde". In: De Plantenwortel in de Landbouw. Ed. by Nederlands Genootschap voor Landbouwwetenschap. S-Gravenhage 1955, pp. 107–118

Borchert, H.: Ein methodischer Beitrag zur Entnahme von Bodenproben in ungestörter Lagerung. Z. Pflanzenernähr. Düng. Bodenkde. *93*, 210–214 (1961)

Bormann, F.H., Graham, B.F.: The occurrence of natural root grafting in eastern white pine, *Pinus strobus* L., and its ecological implications. Ecology *40*, 677–691 (1959)

Bose, R.D., Dixit, P.D.: Studies in Indian barleys. II. The root-system. Indian J. Agric. Sci. *1*, 98–108 (1931)

Bosse, G.: Die Wurzelentwicklung von Apfelklonen und Apfelsämlingen während der ersten drei Standjahre. Erwerbsobstbau *2*, 26–30 (1960)

Bray, J.R.: Root production and the estimation of net productivity. Can. J. Bot. *41*, 65–72 (1963)

Bray, R.A., Hacker, J.B., Byth, D.E.: Root mapping of three tropical pasture species using ^{32}P. Aust. J. Exp. Agric. Anim. Husb. *9*, 445–448 (1969)

Breda, van, N.G.: An improved method in the study of root bisects. S. Afr. J. Sci. *34*, 260–264 (1937)

Breteler, H.G.M., van den Broek, J.M.M.: Drawings on transparent plastic foil as an aid to soil profile descriptions. (Dutch, Engl. Sum.). Boor Spade *17*, 54–63 (1971)

Breviglieri, N.: Studies on the root system of fruit trees and vines in Italy. Rep. 13th Int. Hortic. Congr. 1952. London, 1953. Vol.I, pp. 159–180

Brouwer, R.: Some aspects of the equilibrium between overground and underground plant parts. Instituut voor Biol. en Scheikund. Onderz. van Landbouwgew., Wageningen. Jaarboek 1963, pp. 31–39

Brouwer, R.: Root growth of grasses and cereals. In: The Growth of Cereals and Grasses. Milthorpe, F.L., Ivins, J.D. (eds). London: Butterworth, 1966, pp. 153–166

Brouwer, R.: Beziehungen zwischen Spross- und Wurzelwachstum. Angew. Bot. *41*, 244–254 (1967)

Brouwer, R., De Wit, C.T.: A simulation model of plant growth with special attention to root growth and its consequences. In: Root Growth. Whittington, W.J. (ed). London: Butterworth, 1969, pp. 224–244

Brouwer, R., Jenneskens, P.J., Borggreve, G. J.: Growth responses of shoots and roots to interruptions of the nitrogen supply. Instituut voor Biol. en Scheikund. Onderz. van Landbouwgew., Wageningen. Jaarboek 1961, pp. 29–36

Brown, G.R., Thilenius, J.F.: A low-cost machine for separation of roots from soil material. J. Range Manage. *29*, 506–507 (1976)

Brown, J.H.: Variation in roots of greenhouse grown seedlings of different Scotch pine provenances. Silvae Genet. *18*, 111–117 (1969)

Brückner, E.: "Fraget die Bäume!" Allg. Forstz. *5*, 427–428 (1950)

Bruner, W.E.: Root development of cotton, peanut and tobacco in Central Oklahoma. Proc. Oklahoma Academy of Science for 1931 *12*, 20–37 (1932)

Buchele, W.F.: A power sampler of undisturbed soils. Trans. Am. Soc. Agric. Eng. *4*, 185–187 and 191 (1961)

Bul'Botko, G.V.: The effect of the physical properties of soils on the development of the root system of apples. Sov. Soil Sci. (Washington, Engl. translation from Pochvovedeniye) *5*, 219–224 (1973)

Bunt, A.C., Kulwiec, Z.J.: The effect of container porosity on root environment and plant growth. I. Temperature. Plant Soil *32*, 65–80 (1970)

Burton, G.W.: A comparison of the first year's root production of seven southern grasses established from seed. J. Am. Soc. Agron. *35*, 192–196 (1943)

Burton, G.W.: Role of tracers in root development investigations. In: Atomic Energy and Agriculture. Comar, C.L. (ed). American Assoc. for the Advancement of Science, Washington D.C. Publ. No. 49, 1957, pp. 71–80

Burton, G.W., De Vane, E.H., Carter, R.L.: Root penetration, distribution and activity in Southern grasses measured by yields, drought symptoms and P 32 uptake. Agron. J. *46*, 229–233 (1954)

Butijn, H.J.: Bewortelingsproblemen in de fruitteelt. In: De Plantenwortel in de Landbouw. Ed. by Nederlands Genootschap voor Landbouwwetensch. 'S-Gravenhage 1955, pp. 156–167

Butijn, H.J.: De betekenis van bewortelingsopnamen in de fruitteelt. Meded. Dir. van de Tuinbouw *21*, 622–631 (1958)

Byle, De, N.V.: Detection of functional intraclonal aspen root connections by tracers and excavation. For. Sci. *10*, 386–396 (1964)

Cahoon, G.A., Morton, E.S.: An apparatus for the quantitative separation of plant roots from soil. Proc. Am. Soc. Hortic. Sci. *78*, 593–596 (1961)

Cahoon, G.A., Stolzy, L.H.: Estimating root density and distribution in citrus orchards by the neutron moderation method. Proc. Am. Soc. Hortic. Sci. *74*, 322–327 (1959)

Caldwell, M.M., Camp, L.B.: Belowground productivity of two cool desert communities. Oecologia (Berl.) *17*, 123–130 (1974)

Caldwell, M.M., Fernandez, O.A.: Dynamics of Great Basin shrub root systems. In: Environmental Physiology of Desert Organisms. Hadley, N.F. (ed.). Stroudsburg, Penn. (USA): Dowden, Hutchinson and Ross Inc., 1975, pp. 38–51

Caldwell, M.M., White, R.S., Moore, R.T., Camp, L.B.: Carbon balance, productivity, and water use of cold-winter desert shrub communities dominated by C_3 and C_4 species. Oecologia (Berl.) *29*, 275–300 (1977)

Cannon, W.A.: The Root Habits of Desert Plants. Carnegie Institution of Washington. Publ. No. *131*, Washington D.C. 1911

Canode, C.L., Hiebert, A.J., Russel, T.S., Law, A.G.: Sampling technique for estimation of root production of Kentucky bluegrass. Crop Sci. *17*, 28–30 (1977)

Carley, H.E., Watson, R.D.: A new gravimetric method for estimating root-surface areas. Soil Sci. *102*, 289–291 (1966)

Carlson, C.A.: A core method for determining the amount and extent of small roots. U.S. South. For. Exp. St. Occas. Pap. *135*, 43–47 (1954)

Carlson, C.W.: Problems and techniques in studying plant root systems. Proc. 9th Int. Grassl. Congr. Sao Paulo. 1965, Vol. 2, pp. 1491–1493

Carson, E.W. (ed.): The Plant Root and its Environment. Charlottesville, USA. University Press of Virginia, 1974.

Carter, W.: A method of growing plants in water vapor to facilitate examination of roots. Phytopathology *32*, 623–625 (1942)

Castle, W.S., Krezdorn, A.H.: Soil water use and apparent root efficiencies of citrus trees on four rootstocks. J. Am. Soc. Hortic. Sci. *102*, 403–406 (1977)

Chandler, W.V.: Effects of long-time surface fertilization on rooting depth and habits of oats. Agron. J. *50*, 286 (1958)

Chaudhary, M.R., Prihar, S.S.: Root development and growth response of corn following mulching, cultivation, or interrow compaction. Agron. J. *66*, 350–355 (1974)

Chilvers, G.A.: Tree root pattern in a mixed eucalypt forest. Aust. J. Bot. *20*, 229–234 (1972)

Chloupek, O.: The relationship between electric capacitance and some other parameters of plant roots. Biol. Plant. *14*, 227–230 (1972a)

Chloupek, O.: Der Anteil der Wurzeln an der Ertragsbildung des Winterweizens bei verschiedener Düngung. Z. Acker- Pflanzenbau *136*, 164–169 (1972b)

Chloupek, O.: Die Bewertung des Wurzelsystems von Senfpflanzen auf Grund der dielektrischen Eigenschaften und mit Rücksicht auf den Endertrag. Biol. Plant. *18*, 44–49 (1976a)

Chloupek, O.: The size of the root system of tetraploid red clover and its relation to the chemical composition of the herbage produced. J. Br. Grassl. Soc. *31*, 23–27 (1976b)

Chloupek, O.: Evaluation of the size of a plant's root system using its electrical capacitance. Plant Soil *48*, 525–532 (1977)

Cholick, F.A., Welsh, J.R., Cole, C.V.: Rooting patterns of semi-dwarf and tall winter wheat cultivars under dryland field conditions. Crop Sci. *17*, 637–639 (1977)

Christensen, J.R.: Root studies XI. Raspberry root systems. J. Pomol. Hortic. Sci. *23*, 218–226 (1947)

Claassen, N., Barber, S.A.: A method for characterizing the relation between nutrient concentration and flux into roots of intact plants. Plant Physiol. *54*, 564–568 (1974)
Clausen, D.: Oberirdische Pflanzenmasse und Wurzelgewicht. Fortschr. Landwirtsch. *4*, 277–280 (1929)
Clayton, M.F., Lamberton, J.A.: A study of root exudates by the fog-box technique. Aust. J. Biol. Sci. *17*, 855–866 (1964)
Cockroft, B., Wallbrink, J.C.: Root distribution of orchard trees. Aust. J. Agric. Res. *17*, 49–54 (1966)
Coetzee, J.A., Page, M.I., Meredith, D.: Root studies in highveld grassland communities. S. Afr. J. Sci. *42*, 105–118 (1946)
Coker, E.G.: The root development of black currants under straw mulch and clean cultivation. J. Hortic. Sci. *33*, 21–28 (1958a)
Coker, E.G.: Root Studies XII. Root systems of apple on Malling rootstocks on five soil series. J. Hortic. Sci. *33*, 71–79 (1958b)
Coker, E.G.: Root development of apple trees in grass and clean cultivation. J. Hortic. Sci. *34*, 111–121 (1959)
Conrad, J.P., Veihmeyer, F.J.: Root development and soil moisture. Hilgardia *4*, 113–134 (1929)
Cook, C.W.: A study of the roots of *Bromus inermis* in relation to drought resistance. Ecology *24*, 169–182 (1943)
Cook, R.L., Millar, C.E.: Some techniques which help to make greenhouse investigations comparable with field plot experiments. Proc. Soil Sci. Soc. Am. *11*, 298–304 (1946)
Cooper, A.J.: Root Temperature and Plant Growth. A Review. Commonwealth Agricultural Bureaux, Farnham Royal, Slough, England, 1973
Copes, D.L.: Influence of rooting media on root structure and rooting percentage of Douglas-fir cuttings. Silvae Genetica *26*, 102–106 (1977)
Cormack, R.G.H.: The effect of environmental factors on the development of root hairs in *Phleum pratense* and *Sporobolus cryptandrus*. Am. J. Bot. *31*, 443–449 (1944)
Cornforth, I.S.: Relationships between soil volume used by roots and nutrient accessibility. J. Soil Sci. *19*, 291–301 (1968)
Coupland, R.T., Johnson, R.E.: Rooting characteristics of native grassland species in Saskatchewan. J. Ecol. *53*, 475–507 (1965)
Cox, T.L., Harris, W.F., Ausmus, B.S., Edwards, N.T.: The role of roots in biogeochemical cycles in an eastern deciduous forest. Pedobiologia *18*, 264–271 (1978)
Craig, J.: Selection for root strength in maize. Trop. Agric. *45*, 343–345 (1968)
Crapo, N.L., Coleman, D.C.: Root distribution and respiration in a Carolina old field. Oikos *23*, 137–139 (1972)
Crider, F.J.: Root studies of citrus trees with practical applications. Citrus Leaves 7, No. 4, 1–3, 27–30 (1927)
Cullen, P.W., Turner, A.K., Wilson, J.H.: The effect of irrigation depth on root growth of some pasture species. Plant Soil *37*, 345–352 (1972)
Cullinan, F.P.: Root development of the apple as affected by cultural practices. Proc. Am. Soc. Hortic. Sci. *18*, 197–203 (1921)
Dahlman, R.C.: Root production and turnover of carbon in the root-soil matrix of a grassland ecosystem. In: Methods of Productivity Studies in Root Systems and Rhizosphere Organisms. Int. Symp. USSR 1968. Ed. by USSR Academy of Sciences, Leningrad: Nauka, 1968, pp. 11–21
Danielson, R.E.: Root systems in relation to irrigation. Agronomy *11*, 390–424 (1967)
Dart, P.J.: Legume root nodule initiation and development. In: The Development and Function of Roots. Torrey, J.G., Clarkson, D.T. (eds.). London: Academic Press, 1975, pp. 467–506
Davis, C.H.: Absorption of soil moisture by maize roots. Bot. Gaz. *101*, 791–805 (1940)
Davis, D.B., Runge, E.C.A.: Comparison of rooting in soil parent materials using undisturbed soil cores. Agron. J. *61*, 518–521 (1969)
Davis, R.G., Johnson, W.C.: A growth sensor for sugarbeet roots. Agron. J. *62*, 837–838 (1970)

Davis, R.G., Wiese, A.F., Pafford, J.L.: Root moisture extraction profiles of various weeds. Weeds *13*, 98–100 (1965)

Davis, R.G., Johnson, W.C., Wood, F.O.: Weed root profiles. Agron. J. *59*, 555–556 (1967)

Day, M.W.: The root system of aspen. Am. Midl. Nat. *32*, 502–509 (1944)

Dean, A.L.: Root observation boxes. Phytopathology *19*, 407–412 (1929)

Derera, N.F., Marshall, D.R., Balaam, L.N.: Genetic variability in root development in relation to drought tolerance in spring wheats. Exp. Agric. *5*, 327–337 (1969)

Derick, R.A., Hamilton, D.G.: Root development in oat varieties. Sci. Agric. *22*, 503–508 (1942)

Dittmer, H.J.: A quantitative study of roots and root hairs of a winter rye plant *(Secale cereale)*. Am. J. Bot. *24*, 417–420 (1937)

Dittmer, H.J.: A quantitative study of the subterranean members of three field grasses. Am. J. Bot. *25*, 654–657 (1938)

Dittmer, H.J.: A comparative study of the number and length of roots produced in nineteen Angiosperm species. Bot. Gaz. *109*, 354–357 (1948)

Dittmer, H.J.: Root hair variations in plant species. Am. J. Bot. *36*, 152–155 (1949)

Dittmer, H.J.: A study of the root systems of certain sand dune plants in New Mexico. Ecology *40*, 265–273 (1959a)

Dittmer, H.J.: A method to determine the length of individual roots. Bull. Torrey Bot. Club *86*, 59–61 (1959b)

Dittmer, H.J., Talley, B.P.: Gross morphology of tap roots of desert cucurbits. Bot. Gaz. *125*, 121–126 (1964)

Dobrynin, G.M.: The study of root and shoot systems of grasses by replanting them after excavation. In: Methods of Productivity Studies in Root Systems and Rhizosphere Organisms. Int. Symp. USSR 1968. Ed. by USSR Academy of Sciences, Leningrad: Nauka, 1968, pp. 21–24

Dominik, T.: Key to ectotrophic mycorrhiza. Folia Forest Polon. Ser. A *15*, 309–328 (1969)

Doss, B.D., Ashley, D.A., Bennett, O.L.: Effect of soil moisture regime on root distribution of warm season forage species. Agron. J. *52*, 569–572 (1960)

Drew, M.C., Saker, L.R.: Relationship between root number determined in the field on the horizontal faces of soil cores, and the lengths and weights of roots extracted from the soil. Agric. Res. Counc. Letcombe Laboratory, Wantage, England. Annu. Rep. 1976, pp. 34–35 (1977)

Du Hamel Du Monceau, H.: Naturgeschichte der Bäume. Teil I und II. Nürnberg: Winterschmidt, 1764/65

Duncan, W.G., Ohlrogge, A.J.: Principles of nutrient uptake from fertilizer bands. II. Root development in the band. Agron. J. *50*, 605–608 (1958)

Dunham, C.W.: Use of methylene blue to evaluate rooting of cuttings. Proc. Am. Soc. Hortic. Sci. *72*, 450–453 (1958)

Dunn, J.H., Engel, R.E.: Rooting ability of Merion Kentucky bluegrass sod grown on mineral and muck soil. Agron. J. *62*, 517–520 (1970)

Dunn, J.H., Engel, R.E.: Effect of defoliation and root-pruning on early growth from Merion Kentucky bluegrass sods and seedlings. Agron. J. *63*, 659–663 (1971)

Durrant, M.J., Love, B.J.G., Messem, A.B., Draycott, A.P.: Growth of crop roots in relation to soil moisture extraction. Ann. Appl. Biol. *74*, 387–394 (1973)

Eavis, B.W.: Soil physical conditions affecting seedling root growth. I. Mechanical impedance, aeration and moisture availability as influenced by bulk density and moisture levels in a sandy loam soil. Plant Soil *36*, 613–622 (1972)

Ehlers, W.: Einfluss von Wassergehalt, Struktur und Wurzeldichte auf die Wasseraufnahme von Weizen auf Löss-Parabraunerde. Mitt. Dtsch. Bodenkundl. Ges. *22*, 141–156 (1975)

Ehwald, E., Hausdörfer, H.D., Kundler, P., Vetterlein, E.: Standortkundliche Untersuchungen an diluvialen Sandböden in der Lehroberförsterei Finowtal. Arch. Forstwes. *4*, 481–500 (1955)

Ellern, S.J.: A re-usable experimental container for root studies. Isr. J. Agric. Res. *18*, 135–136 (1968)

Ellern, S.J., Harper, J.L., Sagar, G.R.: A comparative study of the distribution of the roots of *Avena fatua* and *A. strigosa* in mixed stands using a ^{14}C-labelling technique. J. Ecol. *58*, 865–868 (1970)
Ellis, F.B., Barnes, B.T.: Estimation of the distribution of living roots of plants under field conditions. Plant Soil *39*, 81–91 (1973)
Ellis, F.B., Barnes, B.T.: Effects of cultivation systems on the distribution of cereal roots in clay or clay loam soils. Agric. Res. Counc. Letcombe Laboratory, Wantage, England. Annu. Rep. 1976, pp. 35–37 (1977)
Engler, A.: Untersuchungen über das Wurzelwachstum der Holzarten. Mitt. Schweiz. Centralanst. Forstl. Versuchswes. *7*, 247–317 (1903)
Estey, R.H.: A transparent root-observation cell. Can. J. Plant Sci. *48*, 547–549 (1968)
Estey, R.H.: Modifications to extend the usefulness of root-observation cells. Can. J. Plant Sci. *50*, 209–210 (1970)
Evans, H.: Studies on the absorbing surface of suger-cane root systems. I. Method of study with some preliminary results. Ann. Bot. N.S. *2*, 159–182 (1938)
Evans, P.S.: Root growth of *Lolium perenne* I. 1. Effect of plant age, seed weight, and nutrient concentration on root weight, length, and number of apices. N. Z. J. Bot. *8*, 344–356 (1970)
Evans, P.S.: Comparative root morphology of some pasture grasses and clovers. New Z. J. Agric. Res. *20*, 331–335 (1977)
Evans, T.R.: A new technique for placement of radioactive solutions in root studies and a study of root growth of *Paspalum commersonii*. Aust. J. Exp. Agric. Anim. Husb. *7*, 447–452 (1967)
Evdokimova, T.I., Grishina, L.A.: Productivity of root systems of herbaceous vegetation on flood plain meadows and methods for its study. In: Methods of Productivity Studies in Root Systems and Rhizosphere Organisms. Int. Symp. USSR 1968. Ed. by USSR Academy of Sciences, Leningrad: Nauka, 1968, pp. 24–27
Evetts, L.L., Burnside, O.C.: Early root and shoot development of nine plant species. Weed Sci. *21*, 289–291 (1973)
Farris, N.F.: Root habits of certain crop plants as observed in the humid soils of New Jersey. Soil Sci. *38*, 87–111 (1934)
Fayle, D.C.F.: Radial Growth in Tree Roots. University of Toronto. Faculty of Forestry. Tech. Rep. No. 9 (1968)
Fehrenbacher, J.B., Alexander, J.D.: A method for studying corn root distribution using a soil-core sampling machine and a shaker-type washer. Agron. J. *47*, 468–472 (1955)
Fehrenbacher, J.B., Rust, R.H.: Corn root penetration in soils derived from various textures of Wisconsin-age glacial till. Soil Sci. *82*, 369–378 (1956)
Fehrenbacher, J.B., Snider, H.J.: Corn root penetration in Muscatine, Elliott, and Cisne Soils. Soil Sci. *77*, 281–291 (1954)
Fehrenbacher, J.B., Johnson, P.R., Odell, R.T., Johnson, P.E.: Root penetration and development of some farm crops as related to soil physical and chemical properties. 7th Int. Congr. Soil Sci. Madison, Wisc. USA. Transactions. 1960, Vol. III, pp. 243–252
Fehrenbacher, J.B., Ray, B.W., Edwards, W.M.: Rooting volume of corn and alfalfa in shale-influenced soils in Northwestern Illinois. Proc. Soil Sci. Soc. Am. *29*, 591–594 (1965)
Fehrenbacher, J.B., Ray, B.W., Alexander, J.D.: Root development of corn, soybeans, wheat, and meadow in some contrasting Illinois soils. Ill. Res. *9*, No. 2, 3–5 (1967)
Fehrenbacher, J.B., Ray, B.W., Alexander, J. D.: How soils affect root growth. Crops Soils *21*, No. 4, 14–18 (1969)
Fehrenbacher, J.B., Ray, B.W., Alexander, J.D., Kleiss, H.J.: Root penetration in soils with contrasting textural layers. Ill. Res. *15*, No. 3, 5–7 (1973)
Fergedal, L.: A method for washing-out and investigating root systems. (Swed., Engl. Sum.). Grundförbättring *20*, 53–60 (1967)
Fernandez, O.A., Caldwell, M.M.: Phenology and dynamics of root growth of three cool semi-desert shrubs under field conditions. J. Ecol. *63*, 703–714 (1975)
Fernandez, O.A., Caldwell, M.M.: Root growth of *Atriplex confertifolia* under field conditions. In: The Belowground Ecosystem: A Synthesis of Plant-Associated Processes.

Marshall, J.K. (ed.). Range Science Department Science Series No. 26, Colorado State University, Fort Collins, USA, 1977, pp. 117–121
Ferrill, M.D., Woods, F.W.: Root extension in a longleaf pine plantation. Ecology *47*, 97–102 (1966)
Finney, J.R., Knight, B.A.G.: The effect of soil physical conditions produced by various cultivation systems on the root development of winter wheat. J. Agric. Sci. *80*, 435–442 (1973)
Fisher, C.: The new Quantimet 720. Microscope *19*, 1–20 (1971)
Fitter, A.H.: Effects of nutrient supply and competition from other species on root growth of *Lolium perenne* in soil. Plant Soil *45*, 177–189 (1976)
Fitzpatrick, E.G., Rose, L.E.: A study of root distribution in prairie claypan and associated friable soils. Bulletin. Am. Soil Surv. Assoc. *17*, 136–145 (1936)
Fleming, A.L., Foy, C.D.: Root structure reflects differential aluminium tolerance in wheat varieties. Agron. J. *60*, 172–176 (1968)
Flocker, W.J., Timm, H.: Plant growth and root distribution in layered sand columns. Agron. J. *61* 530–534 (1969)
Follet, R.F., Allmaras, R.R., Reichman, G.A.: Distribution of corn roots in sandy soil with a declining water table. Agron. J. *66*, 288–292 (1974)
Foos, K.: Eine Wurzelküvette für die kontinuierliche quantitative Erfassung des Wurzelwachstums junger Bäume bei verschiedenen Bedingungen im Wurzelraum. Angew. Bot. *49*, 179–185 (1975)
Ford, E.D., Deans, J.D.: Growth of a Sitka spruce plantation: Spatial distribution and seasonal fluctuations of length, weights, and carbohydrate concentrations of fine roots. Plant Soil *47*, 463–485 (1977)
Ford, H.W.: Thickness of a subsoil organic layer in relation to tree size and root distribution of citrus. Proc. Am. Soc. Hortic. Sci. *82*, 177–179 (1963)
Ford, H.W., Reuther, W., Smith, P.F.: Effect of nitrogen on root development of Valencia orange trees. Proc. Am. Soc. Hortic. Sci. *70*, 237–244 (1957)
Forde, St.C.M.: Effect of dry season drought on uptake of radioactive phosphorus by surface roots of the oil palm (*Elaeis guineensis* Jacq.). Agron. J. *64*, 622–623 (1972)
Fordham, R.: Observations on the growth of roots and shoots of tea (*Camellia sinensis*, L.) in Southern Malawi. J. Hortic. Sci. *47*, 221–229 (1972)
Fors, L.: Preliminary root chamber studies. Meddelanden från Avdelningen för Ekologisk Botanik Lunds Universitet *1*, No. 1, 1–20 (1973)
Foth, H.D.: Root and top growth of corn. Agron. J. *54*, 49–52 (1962)
Foth, H.D., Pratt, J.N.: Soil-root monoliths: collection and preservation. Agron. J. *51*, 246–247 (1959)
Fox, R.L., Lipps, R.C.: Influence of soil-profile characteristics upon the distribution of roots and nodules of alfalfa and sweetclover. Agron. J. *47*, 361–367 (1955a)
Fox, R.L., Lipps, R.C.: Subirrigation and plant nutrition. I. Alfalfa root distribution and soil properties. Proc. Soil Sci. Soc. Am. *19*, 468–473 (1955b)
Fox, R.L., Lipps, R.C.: A comparison of stable strontium and P^{32} as tracers for estimating alfalfa root activity. Plant Soil *20*, 337–350 (1964)
Fox, R.L., Weaver, J.E., Lipps, R.C.: Influence of certain soil-profile characteristics upon the distribution of root of grasses. Agron. J. *45*, 583–589 (1953)
Foy, C.D., Fleming, A.L., Gerloff, G.C.: Differential aluminium tolerance in two snap bean varieties. Agron. J. *64*, 815–818 (1972)
Fraas, C.: Das Wurzelleben der Kulturpflanzen und die Ertragssteigerung. Berlin: Wiegandt u. Hempel, 2. Aufl., 1872
Frankena, I.H.J., Goedewaagen, M.A.J.: Een vakkenproef over den invloed van verschillende waterstanden op den grasgroei bij drie grondsoorten. Verslagen van Landbouwkundige Onderzoekingen No. 48 (6) A, 407–461 (1942)
Fraser, A.I.: The soil and roots as factors in tree stability. Forestry *35*, 117–127 (1962)
Fraser, A.I., Gardiner, J.B.H.: Rooting and stability in Sitka spruce. Forestry Commission. Bulletin No. 40. London: Her Majesty's Stationary Office, 1967, 28 pp
Fred, E.B., Baldwin, J.L., McCoy, E.: Root Nodule Bacteria and Leguminous Plants. University of Wisconsin Studies No. 5. Madison 1932

Freeman, B.M., Smart, R.E.: A root observation laboratory for studies with grapevines. Am. J. Enol. Vitic. *27*, 36–39 (1976)

Frese, H., Czeratzki, W., Altemüller, H.J.: Über die Wirkung der Verteilung organischer und anorganischer Düngerstoffe im Boden auf das Wurzelwachstum von Zuckerrüben. Z. Pflanzenernähr. Düng. Bodenk. *69*, 198–205 (1955)

Freytag, H.E., Jäger, R.: Zur Verfolgung des Wurzellängenwachstums mit Hilfe von ^{32}P-Injektionen. Arch. Acker- Pflanzenbau Bodenkde. *16*, 521–526 (1972)

Fribourg, H.A.: A rapid method for washing roots. Agron. J. *45*, 334–335 (1953)

Frielinghaus, M., Spitzl, M.: Wirkung der Tieflockerung und Tiefdüngung auf die Durchwurzelung des Unterbodens einer Tieflehm-Fahlerde. Arch. Acker- Pflanzenbau Bodenkde. *20*, 369–380 (1976)

Frischenschlager, B.: Wurzeluntersuchungen bei Apfel, Birne, Zwetschke, Kirsche und Walnuß. Gartenbauwissenschaft *9*, 269–292 (1935)

Fritts, H.C.: Tree Rings and Climate. London-New York-San Francisco: Academic Press, 1976

Fritzsche, R., Nyfeler, A.: Beeinflussung der Entwicklung und Aktivität der Wurzeln von Apfelbäumen durch Bodenpflege- und Bewirtschaftungsmaßnahmen. Schweiz. Landwirtsch. Forsch. *13*, 341–351 (1974)

Fröhlich, H.J.: Die Bodendurchwurzelung seitens verschiedener Gemüsearten. Arch. Gartenbau *4*, 389–417 (1956)

Fröhlich, H.J., Dietze, W.: Untersuchungen über Wurzelentwicklung an Pflanzen der Gattung *Populus*, Sektionen Aigeiros, Leuce and Tacamahaca. Silvae Genet. *19*, 131–142 (1970)

Fruhwirth, C.: Ueber die Ausbildung des Wurzelsystems der Hülsenfrüchte. Forsch. Geb. Agrikulturphy. *18*, 461–479 (1895)

Fryrear, D.W., McCully, W.G.: Development of grass root systems as influenced by soil compaction. J. Range Manage. *25*, 254–257 (1972)

Gäde, H.: Untersuchungen über die Bewurzelungsverhältnisse gelbblühender Lupinen auf leichten Böden. Albrecht-Thaer-Arch. *6*, 359–375 (1962)

Gaiser, R.N.: Root channels and roots in forest soils. Proc. Soil Sci. Soc. Am. *16*, 62–65 (1952)

Gardner, W.R.: Relation of root distribution to water uptake and availability. Agron. J. *56*, 41–45 (1964)

Garin, G.I.: Distribution of roots of certain tree species in two Connecticut soils. Connecticut Agric. Exp. Stn. Bull. No. 454, 97–167 (1942)

Garner, M.R., Telfair, D.: New techniques for the study of restoration of compacted soil. Science *120*, 668–669 (1954)

Garwood, E.A.: Seasonal variation in appearance and growth of grass roots. J. Br. Grassl. Soc. *22*, 121–130 (1967)

Gates, C.T.: Quantitative recovery of root systems in pot experiments. J. Aust. Inst. Agric. Sci. *17*, 152–154 (1951)

Geisler, G.: Die Bedeutung des Wurzelsystems für die Züchtung dürreresistenter Rebenunterlagssorten. Vitis *1*, 14–31 (1957)

Geisler, G.: Untersuchungen zum Einfluß der Bodenluft auf das Wurzelwachstum. Z. Acker- Pflanzenbau *118*, 399–410 (1964)

Geisler, G.: Bodenluft und Pflanzenwachstum unter besonderer Berücksichtigung der Wurzel. Arb. Landwirtsch. Hochsch. Hohenheim *40* (1967) 111 pp

Geisler, G., Maarufi, D.: Untersuchungen zur Bedeutung des Wurzelsystems von Kulturpflanzen. Teil I. Der Einfluß des Bodenwassergehaltes und der Stickstoffdüngung auf Pflanzenwachstum, Wurzelmorphologie, Transpiration und Stickstoffaufnahme. Z. Acker- Pflanzenbau *141*, 211–230 (1975)

Gerdemann, J.W.: Mycorrhizae. In: The Plant Root and its Environment. Carson, E.W. (ed.). Charlottesville: University Press of Virginia, 1974, pp. 205–217

Gerdemann, J.W.: Vesicular-arbuscular mycorrhizae. In: The Development and Function of Roots. Torrey, J.G., Clarkson, D.T. (eds.). London: Academic Press, 1975, pp. 575–591

Gericke, S.: Die Bedeutung der Ernterückstände für den Humushaushalt des Bodens. 1. Teil: Die Ergebnisse bisheriger Untersuchungen. Z. Pflanzenernähr. Düng. Bodenkde. *35*, 229–247 (1945)

Gericke, S.: Die Bedeutung der Ernterückstände für den Humushaushalt des Bodens. 2. Teil: Die Ergebnisse eigener Untersuchungen. Z. Pflanzenernähr. Düng. Bodenkde. *37*, 46–61 (1946)

Gerlagh, M.: A perspex box for microscopic observation of living roots. Neth. J. Plant. Pathol. *72*, 248–249 (1966)

Gerwitz, A., Page, E.R.: Estimation of root distribution in soil by labelling with rubidium-86 and counting with commercially available equipment. Lab. Pract. *22*, 35–36 (1973)

Gerwitz, A., Page, E.R.: An empirical mathematical model to describe plant root systems. J. Appl. Ecol. *11*, 773–782 (1974)

Gier, L.J.: Root systems of bright belt tobacco. Am. J. Bot. *27*, 780–787 (1940)

Giesecke, F.: Der Vegetationsversuch. 2. Der Gefäßversuch und seine Technik. Methodenbuch Band IX. Radebeul u. Berlin: Neumann, 1954

Gifford, G.F.: Aspen root studies on three sites in Northern Utah. Am. Midl. Nat. *75*, 132–141 (1966)

Gile, P.L., Carrero, J.O.: Absorption of nutrients as affected by the number of roots supplied with the nutrient. J. Agric. Res. *9*, 73–95 (1917)

Giskin, M.L., Kohnke, H.: Measuring root responses to soil properties with a single plant. Agron. J. *57*, 96–97 (1965)

Glatzel, K.: Boden- und Wurzelfotografie. Allg. Forstz. *19*, 437–439 (1964)

Gliemeroth, G.: Wasserhaushalt des Bodens in Abhängigkeit von der Wurzelausbildung einiger Kulturpflanzen. Z. Acker- Pflanzenbau *95*, 21–46 (1952)

Gliemeroth, G.: Bearbeitung und Düngung des Unterbodens in ihrer Wirkung auf Wurzelentwicklung, Stoffaufnahme und Pflanzenleistung. Z. Acker- Pflanzenbau *96*, 1–44 (1953a)

Gliemeroth, G.: Beinigkeit und Wurzelwachstum der Zuckerrüben. Zucker *6*, 573–578 (1953b)

Gliemeroth, G.: Untersuchungen über Ausbildung und Leistung der Keim- und Kronenwurzeln bei Sommergetreide. Z. Acker- Pflanzenbau *103*, 1-21 (1957)

Glover, J.: The simultaneous growth of sugarcane roots and tops in relation to soil and climate. Proc. S. Afr. Sugar Technol. Assoc. *41*, 143–159 (1967)

Glover, J.: Further results from Mount Edgecombe root-laboratory. Proc. S. Afr. Sugar Technol. Assoc. *42*, 123–130 (1968a)

Glover, J.: The behaviour of the root system of sugarcane at and after harvest. Proc. S. Afr. Sugar Technol. Assoc. *42*, 133–135 (1968b)

Glover, J.: Soil sterilization and the growth of tops and roots at the Mount Edgecomb root laboratory. Proc. S. Afr. Sugar Technol. Assoc. *44*, 162–167 (1970)

Goedewaagen, M.A.J.: Das Wurzelsystem der Getreidepflanzen bei ungleicher Verteilung der Nährstoffe im Boden. Phosphorsäure *3*, 688–711 (1933)

Goedewaagen, M.A.J.: The relative weight of shoot and root of different crops and its agricultural significance in relation to the amount of phosphate added to the soil. Soil Sci. *44*, 185–202 (1937)

Goedewaagen, M.A.J.: De waterhuishouding van den grond en de wortelontwikkeling. Landbouwkd. Tijdschr. *53*, 118–146 (1941)

Goedewaagen, M.A.J.: Het Wortelstelsel der Landbouwgewassen. 'S-Gravenhage: Alg. Landsdrukkerij, 1942

Goedewaagen, M.A.J.: Beoordeling van de vitaliteit van het wortelstelsel. Landbouwkd. Tijdschr. *66*, 553–555 (1954)

Goedewaagen, M.A.J.: De oecologie van het wortelstelsel der gewassen. In: De Plantenwortel in de Landbouw. Ed. by Nederlands Genootschap voor Landbouwwetenschap. 'S-Gravenhage 1955, pp. 31–68

Goethe, R.: Einige Beobachtungen über Regenwürmer und deren Bedeutung für das Wachsthum der Wurzeln. Jahrb. Nassau. Ver. Naturkd. *48*, 27–34 (1895)

Görbing, J.: Der Einfluß der Phosphorsäure auf die Wurzelentwicklung. Superphosphate *3*, 257–261 (1930)

Görbing, J.: Die Grundlagen der Gare im praktischen Ackerbau. Hannover: Landbuch-Verlag, 2 Bände, o.J., 1948

Göttsche, D.: Verteilung von Feinwurzeln und Mykorrhizen im Bodenprofil eines Buchen- und Fichtenbestandes im Solling. Mitt. Bundesforschungsanst. Forst- Holzwirtsch. Nr. 88, 1–102 (1972)
Goff, E.S.: A study of the roots of certain perennial plants. Wisconcin Agric. Exp. Stn. 14th Annu. Rep. 286–298 (1897)
Gooderham, P.T.: A simple method for the extraction and preservation of an undisturbed root system from a soil. Plant Soil *31*, 201–205 (1969)
Goodman, D., Greenwood, D.J.: Distribution of roots, water and nutrients beneath cabbage grown in the field. J. Sci. Food Agric. *27*, 28–36 (1976)
Gordienko, M.: Über die Beziehungen zwischen Bodenbeschaffenheit und Wurzelgestaltung bei jungen Pflanzen. Landwirtsch. Jahrb. *72*, 125–139 (1930)
Graham, B.F.: Transfer of dye through natural root grafts of *Pinus strobus L.* Ecology *41*, 56–64 (1960)
Graham, B.F., Bormann, F.H.: Natural root grafts. Bot. Rev. *32*, 255–292 (1966)
Greenham, C.G., Cole, D.J.: Diagnosis of death in roots. Nature (London) *164*, 699 (1949)
Gregory, P.J., McGowan, M., Biscoe, P.V., Hunter, B.: Water relations of winter wheat. 1. Growth of the root system. J. Agric. Sci. *91*, 91–102 (1978)
Grimes, D.W., Miller, R.J., Wiley, P.L.: Cotton and corn root development in two field soils of different strength characteristics. Agron. J. *67*, 519–523 (1975)
Grimes, D.W., Sheesley, W.R., Wiley, P.L.: Alfalfa root development and shoot regrowth in compact soil of wheel traffic patterns. Agron. J. *70*, 955–958 (1978)
Grobbelaar, W.P.: Responses of young maize plants to root temperatures. Meded. Landbouwhogesch. Wageningen *63* (5), 1–71 (1963)
Grosse-Brauckmann, E.: Zur Technik des Gefäßversuches: Der Einfluß von Bodenmenge und Gefäßtiefe auf Wachstum und Phosphataufnahme von Hafer. Landwirtsch. Forsch. *25*, 185–190 (1972)
Grosskopf, W.: Bestimmung der charakteristischen Feinwurzel-Intensitäten in ungünstigen Waldbodenprofilen und ihre oekologische Auswertung. Mitt. Bundesforschungsanstalt Forst- Holzwirtsch. Nr. 11, 1–19 (1950)
Guernsey, C.W., Fehrenbacher, J.B., Ray, B.W., Miller, L.B.: Corn yields, root volumes, and soil changes on the Morrow Plots. J. Soil Water Conserv. *24*, 101–104 (1969)
Guiscafré-Arrillaga, J., Gómez, L.A.: Studies of the root system of *Coffea arabica* L. Part I. Environmental condition affecting the distribution of coffee roots in Colosa clay. J. Agric. Univ. Puerto Rico *22*, 227–262 (1938)
Guiscafré-Arrillaga, J., Gómez, L.A.: Studies of the root system of *Coffea arabica* L. Part II. Growth and distribution in Catalina clay soil. J. Agric. Univ. Puerto Rico *24*, 109–117 (1940)
Guiscafré-Arrillaga, J., Gómez, L.A.: Studies of the root system of *Coffea arabica* L. Part III. Growth and distribution of roots of 21-year-old trees in Catalina clay soil. J. Agric. Univ. Puerto Rico *26*, 34–39 (1942)
Günther, E.: Einfluß der Fruchtfolge auf die Bodenfruchtbarkeit. Kühn-Archiv *64*, 34–98 (1951)
Gurr, E.: The Rational Use of Dyes in Biology and General Staining Methods. London: Leonard Hill, 1965
Gwynne, M.D.: Root systems of pineapple plants. East Afr. Agric. For. J. *27*, 204–206 (1962)
Haahr, V.: Moisture extraction from the subsoil by some agricultural crops. Royal Veterinary and Agric. College, Copenhagen, Denmark. Yearb. 1968, 134–145 (1968a)
Haahr, V.: Root development of some cultivated plants grown in a sandy clay loam. (Dan., Engl. Sum.). Tidsskr. Planteavl *72*, 531–538 (1968b)
Haahr, V.: Studies on root development of cereal varieties in the field measured by the plant uptake of soil water and ^{32}P. Eucarpia. Meeting of the Sections Cereal and Physiology. Dijon, France, October 1970, 189–203 (1971)
Haahr, V.: Nuclear methods for detecting root activity. In: Tracer Techniques for Plant Breeding. Ed. by Int. At. En. Ag., Vienne 1975, pp. 57–63
Haahr, V., Knudsen, H., Nielson, G., Sandfoer, J.: Root development of some cultivated plants in light sandy soil. (Dan., Engl. Sum.). Tidsskr. Planteavl *69*, 554–567 (1966)

Haans, J.C.F.M., Houben, J.M.M.Th., van der Sluijs, P.: Properties of hydromorphic sandy soils in relation to root growth. In: Pseudogley and Gley. Transactions of Commissions V and VI, Int. Soc. Soil. Sci. Stuttgart: 1971. Weinheim: 1973: Verlag Chemie, pp. 567–576

Haas, H.J., Rogler, G.A.: A technique for photographing grass roots in situ. Agron. J. *45*, 173 (1953)

Haas, de, P.G., Hein, K.: Über die Beeinflussung des Wachstumsverlaufs von Apfelwurzeln durch verschiedene Baumschnittmaßnahmen und durch Entblätterung. Erwerbsobstbau *15*, 137–141 (1973)

Haas, de, P.G., Jürgensen, C.: Untersuchungen über den Wurzelhabitus von Apfelbäumen auf Hochmoor. Gartenbauwissenschaft *28*, 65–94 (1963)

Hales, S.: Vegetable Staticks. London 1727. Reprint: London: MacDonald, 1961

Hall, N.S., Chandler, W.F., van Bavel, C.H.M., Reid, P.H., Anderson, J.H.: A tracer technique to measure growth and activity of plant root systems. N. C. Agric. Exp. Stn. Tech. Bull. No. 101 (1953) 40 pp

Hamilton, D.G.: Culm, crown and root development in oats as related to lodging. Sci. Agric. *31*, 286–315 (1951)

Hammar, H.E., Smith, C.L., Alben, A.O.: Boron uptake as a criterion of the root spread of pecan trees. Proc. Am. Soc. Hortic. Sci. *62*, 131–134 (1953)

Hammes, J.K., Bartz, J.F.: Root distribution and development of vegetable crops as measured by radioactive phosphorus injection technique. Agron. J. 55, 329–333 (1963)

Hanson, A.A., Juska, F.V.: Winter root activity in Kentucky bluegrass (*Poa pratensis* L.). Agron. J. *53*, 372–374 (1961)

Hanus, H.: Wurzelprofil und Wasserversorgung der Grasnarbe bei verschiedenen Grundwasserständen. Diss. Agric. Bonn, 1962

Hanus, H.: Die Beeinflussung des Wurzelsystems von Weizen durch CCC. Z. Acker- Pflanzenbau *125*, 40–46 (1967)

Hanus, H.: Einfluß verschiedener CCC-Konzentrationen auf die Wurzelentwicklung einzelner Weizensorten. Z. Acker- Pflanzenbau *132*, 59–72 (1970)

Harley, J.L.: The Biology of Mycorrhiza. London: Leonhard Hill, 1969

Harris, F.S.: The effect of soil moisture, plant food, and age on the ratio of tops to roots in plants. J. Am. Soc. Agric. *6*, 65–75 (1914)

Hartmann, H.D.: Wurzelbild und chemische Unkrautbekämpfung bei Kopfkohl. Angew. Bot. *43*, 47–57 (1969)

Hausdörfer, H.-D.: Beiträge zur Kenntnis der Durchwurzelungsverhältnisse unter Buchen und Kiefern auf Sandböden und Lehmböden mit Sanddecken in Nordostdeutschland. Diss. For. der Humboldt-Universität zu Berlin in Eberswalde. Maschinenschr. vervielfältigt, 1959

Havis, L.: Peach tree root distribution. Ecology *19*, 454–462 (1938)

Hays, W.M.: Corn, its habits of root growth, methods of planting and cultivating, notes on ears and stools or suckers. Minnesota Agric. Exp. Stn. Bull. No. 5, 1–33 (1889)

Hays, W.M.: Roots of crops. N. D. Agric. Exp. Stn. Bull. No. 10, 47–49 (1893)

Head, G.C.: Studies of diurnal changes in cherry root growth and nutational movements of apple root tips by time-lapse cinematography. Ann. Bot. N.S. *29*, 219–224 (1965)

Head, G.C.: Estimating seasonal changes in the quantity of white unsuberized root on fruit trees. J. Hortic. Sci. *41*, 197–206 (1966)

Head, G.C.: Effects of seasonal changes in shoot growth on the amount of unsuberized root on apple and plum trees. J. Hortic. Sci. *42*, 169–180 (1967)

Head, G.C.: Seasonal changes in the amounts of white unsuberized root on pear trees on quince rootstock. J. Hortic. Sci. *43*, 49–58 (1968a)

Head, G.C.: Seasonal changes in the diameter of secondarily thickened roots of fruit trees in relation to growth of other parts of the tree. J. Hortic. Sci. *43*, 275–282 (1968b)

Head, G.C.: Studies of growing roots by time-lapse cinematography. Trans. 9th Int. Congr. Soil Sci. Adelaide, Australia, 1968. 1968c, Vol. I, 751–758

Head, G.C.: The effects of fruiting and defoliation on seasonal trends in new root production on apple trees. J. Hortic. Sci. *44*, 175–181 (1969a)

Head, G.C.: The effects of mineral fertilizer on seasonal changes in the amount of white root on apple trees in grass. J. Hortic. Sci. *44*, 183–187 (1969b)
Head, G.C.: Methods for the study of production in root systems. In: Methods of Study in Soil Ecology. Phillipson, J. (ed.). Paris: UNESCO, 1970, pp. 151–157
Head, G.C.: Shedding of roots. In: Shedding of Plant Parts. Kozlowski, T.T. (ed.). New York, London: Academic Press, 1973, pp. 237–293
Heikurainen, L.: Über Veränderungen in den Wurzelverhältnissen der Kiefernbestände auf Moorböden im Laufe des Jahres. Acta For. Fenn. *65*, *2*, 1–70 (1955a)
Heikurainen, L.: Der Wurzelaufbau der Kiefernbestände auf Reisermoorböden und seine Beeinflussung durch die Entwässerung. (Finn., Germ. Sum.). Acta For. Fenn. *65*, *3*, 1–85 (1955b)
Heinisch, O.: Die Bedeutung der Keimwurzelanzahl der Getreidearten für den Saatgutwert. Z. Pflanzenzücht. *22*, 209–232 (1938)
Hellmers, H., Horton, J.S., Juhren, G., O'Keefe, J.: Root systems of some chaparral plants in Southern California. Ecology *36*, 667–678 (1955)
Hellriegel, H.: Beiträge zu den naturwissenschaftlichen Grundlagen des Ackerbaus mit besonderer Berücksichtigung der agrikultur-chemischen Methode der Sandkultur. Braunschweig: F.Vieweg u. Sohn, 1883
Hendrickson, A.H., Veihmeyer, F.J.: Influence of dry soil on root extension. Plant Physiol. *6*, 567–576 (1931)
Hendrix, J.W., Lloyd, E.H.: An improved Washington State University mist chamber for root disease and root research. Wash. Agric. Exp. Stn. Bull. No. 700, 1–5 (1968)
Hensen, V.: Die Wurzeln in den tieferen Bodenschichten. Jahrb. Dtsch. Landwirtschaftsges. *7*, 84–96 (1892)
Hentrich, W.: Untersuchungen über das Wurzelwachstum von Getreidesorten und -arten unter verschiedenen Wachstumsbedingungen. Züchter *36*, 25–36 (1966)
Hess, N.: Beobachtungen über das Wurzelsystem bei Wintergerste. Bodenkultur *3*, 211–214 (1949)
Hesselink, E.: Einiges über die Wurzelentwicklung der Gemeinen Kiefer (*Pinus silvestris* L.) und der Österreichischen Schwarzkiefer (*Pinus laricio austriaca* Endl.). Meded. Rijksboschbouwproefstn. Deel II, 187–278 (1926)
Hewitt, E.J.: Sand and Water Culture Methods used in the Study of Plant Nutrition. Commonwealth Agric. Bur., Farnham Royal, 1966
Heyl, W.: Der Einfluß der Bodenstrukturerkrankungen auf die Wurzelentwicklung landwirtschaftlicher Kulturpflanzen. Kühn-Archiv *56*, 215–245 (1942)
Hignet, C.T.: A method for sampling and measuring cereal roots. J. Aust. Inst. Agric. Sci. *42*, 127–129 (1976)
Hilf, H.H.: Wurzelstudien an Waldbäumen. Hannover: M. H. Schaper, 1927
Hilkenbäumer, F.: Sproß- und Wurzelkronenentwicklung verschiedener Obstarten während der ersten sechs Standjahre auf Lehmboden. Erwerbsobstbau *1*, 127–132 (1959)
Hillel, D., Talpaz, H., van Keulen, H.: A macroscopic-scale model of water uptake by a nonuniform root system and of water and salt movement in the soil profile. Soil Sci. *121*, 242–255 (1976)
Hilton, R.J., Khatamian, H.: Diurnal variation in elongation rates of roots of woody plants. Can. J. Plant Sci. *53*, 699–700 (1973)
Hilton, R.J., Khatamian, H.: In situ observations of woody plant root growth at the University of Guelph rhizotron. In: Ökologie und Physiologie des Wurzelwachstums. II. Int. Symp. 1971. Hoffmann, G. (ed.). Berlin: Akademie-Verlag, 1974, pp. 171–186
Hilton, R.J., Mason, G.F.: Responses of *Mugho* pine roots to soil temperature and of *Bolleana* poplar roots to light. Hortscience *6*, 43–45 (1971)
Hilton, R.J., Bhar, D.S., Mason, G.F.: A rhizotron for in situ root growth studies. Can. J. Plant Sci. *49*, 101–104 (1969)
Hocking, D., Mitchell, D.L.: The influences of rooting volume-seedling espacement and substratum density on greenhouse growth of Lodgepole pine, White spruce, and Douglas fir grown in extruded peat cylinders. Can. J. For. Res. *5*, 440–451 (1975)
Hoffmann, A., Kummerow, J.: Root studies in the Chilean Matorral. Oecologia (Berl.) *32*, 57–69 (1978)

Hoffmann, G.: Verlauf der Tiefendurchwurzelung und Feinwurzelbildung bei einigen Baumarten. Arch. Forstwes. *15*, 825–856 (1966a)
Hoffmann, G.: Auswertung von Wurzelwachstumsuntersuchungen für die praktische Behandlung von Forstbaumschulpflanzen. Sozialistische Forstwirtsch. *16*, 143–147 (1966b)
Hoffmann, G.: Wurzel und Sproßwachstumsperiodik der Jungpflanzen von *Quercus robur* L. im Freiland und unter Schattenbelastung. Arch. Forstwes. *16*, 745–749 (1967)
Hoffmann, G.: Beeinflussung des Wurzel- und Sprosswachstums der Robinie (*Robinia pseudoacacia* L.) durch Bodenbeheizung. Arch. Forstwes. *17*, 431–435 (1968)
Hoffmann, G.: Wachstumsrhythmik der Wurzeln und Sproßachsen von Forstgehölzen. Flora *161*, 303–319 (1972)
Hoffmann, G. (ed.): Ökologie und Physiologie des Wurzelwachstums. II. Int. Symp. Potsdam, 1971. Berlin: Akademie-Verlag, 1974a
Hoffmann, G.: Verfahren zur Messung des Wurzelwachstums unter Freilandbedingungen. In: Ökologie und Physiologie des Wurzelwachstums. II. Int. Symp. 1971. Ed. by G. Hoffmann, Berlin: Akademie-Verlag, 1974b, pp. 161–169
Holbert, J. R., Koehler, B.: Anchorage and extent of corn root systems. J. Agric. Res. *27*, 71–78 (1924)
Holch, A.E., Hertel, E.W., Oakes, W.O., Whitwell, H.H.: Root habits of certain plants of the foothill and alpine belts of Rocky Mountain National Park. Ecol. Monogr. *11*, 327–345 (1941)
Holt, E.C., Fisher, F.L.: Root development of Coastal Bermudagrass with high nitrogen fertilization. Agron. J. *52*, 593–596 (1960)
Homeyer, B., Labenski, K.O., Meyer, B., Thormann, A.: Herstellung von Lysimetern mit Böden in natürlicher Lagerung (Monolith-Lysimeter) als Durchlauf-, Unterdruck- oder Grundwasserlysimeter. Zeitschr. Pflanzenern. u. Bodenkunde *136*, 242–245 (1973)
Hosäus, A.: Ueber das Verhältnis der Wurzeln zu den oberirdischen Pflanzenorganen. Fühlings Landwirtsch. Ztg. *21*, 29–37 (1872)
Hösslin, von, R.: Ein Beitrag zur Kenntnis des Wurzelwachstums einiger Gemüsearten unter natürlichen Verhältnissen. In: Festschrift zum 25jährigen Bestehen des gärtnerischen Hochschulstudiums in Deutschland. Hrsg. vom Bund der Diplomgärtner. Berlin 1954, pp. 47–62
Houben, J.M.M.Th.: Effect van diepe grondbewerking op de beworteling in het tracé van een pijpleiding. Cultuurtechn. Tijdschr. *12*, 21–27 (1972)
Houben, J.M.M.Th.: Wortelontwikkeling en bodemgesteldheid. Bedrijfsontwikkeling *5*, 141–148 (1974)
Hough, W.A., Woods, F.W., McCormack, M.L.: Root extension of individual trees in surface soils of a natural longleaf pine-turkey oak stand. For. Sci. *11*, 223–242 (1965)
Howard, A., Howard, G.L.C.: Some methods suitable for the study of root development. Agric. J. India *13*, Special Indian Science Congress Number 1918, 36–39 (1918)
Howland, P., Griffith, A.L.: The root development of transplants after planting in the field. Emp. For. Rev. *40*, 66–70 (1961)
Huck, M.G.: Root distribution and water uptake patterns. In: The Belowground Ecosystem: A Synthesis of Plant-Associated Processes. Marshall, J.K., (ed.). Range Science Department Science Series No. 26, Colorado State University, Fort Collins, USA, 1977, pp. 215–226
Huck, M.G., Klepper, B., Taylor, H.M.: Diurnal variations in root diameter. Plant Physiol. *45*, 529–530 (1970)
Hunt, R.: Further observations on root-shoot equilibria in perennial ryegrass (*Lolium perenne* L.). Ann. Bot. *39*, 745–755 (1975)
Hunt, R.: Significant relationships in the analysis of root-shoot equilibria. Ann. Bot. *40*, 895–897 (1976)
Hunt, R., Stribley, D.P., Read, D.J.: Root-shoot equilibria in cranberry (*Vaccinium macrocarpon* Ait.). Ann. Bot. *39*, 807–810 (1975)
Hunter, A.S., Kelley, O.J.: A new technique for studying the absorption of moisture and nutrients from soil by plant roots. Soil Sci. *62*, 441–450 (1946a)

Hunter, A.S., Kelley, O. J.: The extension of plant roots into dry soil. Plant Physiol. *21*, 445–451 (1946b)

Hurd, E.A.: Root study of three wheat varieties and their resistance to drought and damage by soil cracking. Can. J. Plant. Sci. *44*, 240–248 (1964)

Hurd, E.A.: Growth of roots of seven varieties of spring wheat at high and low moisture levels. Agron. J. *60*, 201–205 (1968)

Hurd, E.A.: Phenotype and drought tolerance in wheat. Agric. Meteorol. *14*, 39–55 (1974)

Hurd, E.A., Spratt, E.D.: Root patterns in crops as related to water and nutrient uptake. In: Physiological Aspects of Dryland Farming, ed. by U.S. Gupta. New Dehli: Oxford and IBH Publishing Co. 1975, pp. 167–235

Huttel, C.: Root distribution and biomass in three ivory coast rain forest plots. In: Tropical Ecological Systems. Golley, B., Medina, E. (eds.). (Ecological Studies *11*). New York-Berlin-Heidelberg: Springer, 1975, pp. 123–130

Huxley, P.A., Turk, A.: A new root observation laboratory for coffee. Kenya Coffee *32*, (376), 156–159, April 1967

Huxley, P.A., Turk, A.: Preliminary investigations with *Arabica* coffee in a root observation laboratory in Kenya. East Afr. Agric. For. J. *40*, 300–312 (1975)

Huxley, P.A., Patel, R.Z., Kabaara, A.M.: ^{86}RbCl as a tracer for root studies in Kikuyu red loam. East Afr. Agric. For. J. *35*, 340–345 (1970)

Huxley, P.A., Patel, R.Z., Kabaara, A.M., Mitchell, H.W.: Tracer studies with ^{32}P on the distribution of functional roots of *Arabica* coffee in Kenya. Ann. Appl. Biol. *77*, 159–180 (1974)

Innis, G.: Dynamic simulation of the belowground ecosystem. In: The Belowground Ecosystem: A Synthesis of Plant Associated Processes. Marshall, J.K., (ed.). Range Science Department Science Series No. 26, Colorado State University, Fort Collins, USA, 1977, pp. 329–334

International A.E.A.: Root Activity Patterns of some Tree Crops. Ed. by the International Atomic Energy Agency, Vienna. Tech. Rep. Series No. 170, 1975, 154 pp

Jacobs, E., Atsmon, D., Kafkafi, U.: A convenient method of placing radioactive substances in soil for studies of root development. Agron. J. *62*, 303–304 (1970)

Jacobs, H.L.: Windows installed underground assist Davey research workers to gather facts about root growth. Davey Bull. (Kent, Ohio) *19*, No. 10, 4–5 (1931)

Jacques, W.A.: A new type of root sampler. N. Z. J. Sci. Technol. Sect. A, *19*, 267–270 (1937a)

Jacques, W.A.: The effect of different rates of defoliation on the root development of certain grasses. N. Z. J. Sci. Technol. Sect. A, *19*, 441–450 (1937b)

Jacques, W.A.: Root-development in some common New Zealand pasture plants. I. Perennial rye-grass (*Lolium perenne*). N.Z. J. Sci. Technol. Sect. A, *22*, 237–247 (1941)

Jacques, W.A.: Root development in some common New Zealand pasture plants. IV. A method of root separation. N. Z. J. Sci. Technol. Sect. A, *26*, 367–371 (1945)

Jacques, W.A., Schwass, R.H.: Root development in some common New Zealand pasture plants. VII. Seasonal root replacement in perennial ryegrass (*Lolium perenne*), Italian ryegrass (*L. multiflorum*) and tall fescue (*Festuca arundinacea*). N. Z. J. Sci. Technol. Sect. A, *37*, 569 –583 (1956)

Jahn, R., Scheffold, K., Hauck, U.: Wurzeluntersuchungen an Waldbäumen in Baden-Württemberg. Schriftenr. Landesforstverwalt. Baden-Württemberg *33*, (1971) 120 pp

Jean, F.C., Weaver, J.E.: Root Behavior and Crop Yield under Irrigation. Carnegie Institution of Washington, Publ. No. 357, Washington D.C., 1924

Joachim, H.F.: Untersuchungen über die Wurzelausbildung der Pappel und die Standortsansprüche von Pappelsorten. Deutsche Akademie der Landwirtschaftswissenschaften zu Berlin. Wiss. Abh. 7, (1953) 205 pp.

Johanson, N.G., Muzik, T.J.: Some effects of 2,4-D on wheat yield and root growth. Bot. Gaz. *122*, 188–194 (1961)

Johnson, W.C., Davis, R.G.: Growth patterns of irrigated sugarbeet roots and tops. Agron. J. *63*, 649–652 (1971)

Jones, L.H., Haskins, H.D.: Distribution of roots in porous and nonporous plant containers. Plant Physiol. *10*, 511–519 (1935)

Jutras, P.J., Tarjan, A.C.: A novel hydraulic soil auger-screen shaker unit for the collection of root samples. Proc. Soil Crop. Soc. Florida *24*, 154–158 (1964)

Kähäri, J., Elonen, P.: Effect of placement of fertilizer and sprinkler irrigation on the development of spring cereals on the basis of root investigations. J. Sci. Agric. Soc. Finland *41*, 89–104 (1969)

Kaigama, B.K., Teare, J.D., Stone, L.R., Powers, W.L.: Root and top growth of irrigated and nonirrigated grain sorghum. Crop Sci. *17*, 555–559 (1977)

Kalela, E.K.: Über Veränderungen in den Wurzelverhältnissen der Kiefernbestände im Laufe der Vegetationsperiode. Acta For. Fenn. *65,1*, 1–42 (1955)

Kampe, K.: Studien über Bewurzelungsstärke und Wurzeleindringungsvermögen verschiedener Kulturpflanzen. Wissenschaftl. Arch. Landwirtsch. Abt. A Pflanzenbau *2*, 1–48 (1929)

Kar, S., Varade, S.B.: Influence of mechanical impedance on rice seedling root growth. Agron. J. *64*, 80–81 (1972)

Karizumi, N.: Studies on the Form and Distribution Habit of Tree Root. (Jap., Engl. Sum.). Gov. For. Exp. Stn. Tokyo, Jpn., Bull. No. *94* (1957) 205 pp

Karizumi, N.: Estimation of root biomass in forests by soil block sampling. In: Methods of Productivity Studies in Root Systems and Rhizosphere Organisms. Int. Symp. USSR 1968. Ed. by USSR Academy of Sciences, Leningrad: Nauka, 1968, pp. 79–86

Katyal, J.C., Subbiah, B.V.: Root-distribution pattern of some wheat varieties. Indian J. Agric. Sci. *41*, 786–790 (1971)

Kaufmann, M.R.: Water relations of pine seedlings in relation to root and shoot growth. Plant Physiol. *43*, 281–288 (1968)

Kaufmann, M.R., Boswell, S.B., Lewis, L.N.: Effect of tree spacing on root distribution of 9-year old "Washington" navel oranges. J. Am. Soc. Hortic. Sci. *97*, 204–206 (1972)

Kausch, W.: Der Einfluss von edaphischen und klimatischen Faktoren auf die Ausbildung des Wurzelwerkes der Pflanzen unter besonderer Berücksichtigung einiger algerischer Wüstenpflanzen. Habil. Schr. Tech. Hochsch. Darmstadt, 1959

Kausch, W.: Die Primärwurzel von *Zea mays* L. Planta *73*, 328–332 (1967)

Kauter, A.: Beiträge zur Kenntnis des Wurzelwachstums der Gräser. Ber. Schweiz. Bot. Ges. *42*, 37–108 (1933)

Kawatake, M., Nishimura, G., Shimura, K., Ishida, R.: A soil-core sampling machine and a fluctuating washer. (Jap., Engl. Sum.). Bull. Tokai-Kinki Nat. Agric. Exp. Stn. Jpn. *11*, 66–70 (1964)

Keil, G.: Das Wurzelwerk von *Taraxacum officinale* Weber. Beih. Bot. Centralbl. Abt. A, *60*, 57–96 (1941)

Kelley, O.J., Hardman, J.A., Jennings, D.S.: A soil-sampling machine for obtaining two-, three- and four-inch diameter cores of undisturbed soil to a depth of six feet. Proc. Soil Sci. Soc. Am. *12*, 85–87 (1947)

Kemmer, E.: Beobachtungen an Wurzelkörpern von Apfelgehölzen. Züchter *26*, 1–12 (1956)

Kemmer, E.: Das Formenspiel der Einzelwurzeln bei Apfelgehölzen. Erwerbsobstbau *4*, 208–212, 224–228 (1962)

Kemmer, E.: Reizgänge im Wurzelraum von Apfelbäumen. Erwerbsobstbau *5*, 93–96 (1963a)

Kemmer, E.: Das Wuchsbild der Wurzelkörper bei Apfelbäumen unter verschiedenen Voraussetzungen. Erwerbsobstbau *5*, 105–108, 128–130 (1963b)

Kemmer, E.: Nachbauversuche in Apfelbeständen unter besonderer Berücksichtigung des Wurzelverhaltens. Obstbau *82*, 137–139, 158–160 (1963c)

Kemmer, E.: Wurzelkörper von Apfelbäumen unter dem Einfluß eines jeweils vorherrschenden Faktors. Züchter *34*, 59–67 (1964)

Kemp, G.A.: Water bath for germination and root development studies under various temperature combinations of soil and ambient air. Can. J. Plant Sci. *52*, 677–679 (1972)

Kemph, G.S.: Measuring fibrous roots with leaf area meter. J. Range Manage. *29*, 85–86 (1976)

Kendall, W.A., Leath, K.T.: Slant-board culture methods for root observations of red clover. Crop Sci. *14*, 317–320 (1974)

Keresztesi, B.: Morphological characteristic of the *Robinia* root system on different sites of the Great Hungarian Plain. In: Methods of Productivity Studies in Root Systems and Rhizosphere Organisms. Int. Symp. USSR 1968. Ed. by USSR Academy of Sciences, Leningrad: Nauka, 1968, pp. 86–96

Kern, K.G., Moll, W., Braun, H.J.: Wurzeluntersuchungen in Rein- und Mischbeständen des Hochschwarzwaldes (Vfl. Todtmoos 2/I-IV). Allg. Forst- Jagdztg. *132*, 241–260 (1961)

Khasawneh, F. E., Copeland, J.P.: Cotton root growth and uptake of nutrients: Relation of phosphorus uptake to quantity, intensity, and buffering capacity. Proc. Soil. Sci. Soc. Am. *37*, 250–254 (1973)

Kiesselbach. T.A., Weihing, R.M.: The comparative root development of selfed lines of corn and their F_1 and F_2 hybrids. Journal Am. Soc. Agron. *27*, 538–541 (1935)

King, F.H.: Natural distribution of roots in field soils. Wisconsin Agricultural Experiment Station. 9th Annu. Rep. 112–120 (1892)

King, F.H.: The natural distribution of roots in field soils. Wisconsin Agricultural Experiment Station 10th Annu. Rep. 160–164 (1893)

King, J.W., Beard, J.B.: Measuring rooting of sodded turfs. Agron. J. *61*, 497–498 (1969)

Kinman, C.F.: A preliminary report on root growth studies with some orchard trees. Proc. Am. Soc. Hortic. Sci. *29*, 220–224 (1932)

Kira, T., Ogawa, H.: Indirect estimation of root biomass increment in trees. In: Methods of Productivity Studies in Root Systems and Rhizosphere Organisms. Int. Symp. USSR 1968. Ed. by USSR Academy of Sciences, Leningrad: Nauka, 1968, pp. 96–101

Kirby, E.J.M., Rackham, O.: A note on the root growth of barley. J. Appl. Ecol. *8*, 919–924 (1971)

Kittock, D.L., Patterson, J.K.: Measurement of relative root penetration of grass seedlings. Agron. J. *51*, 512 (1959)

Klapp, E.: Über die Wurzelverbreitung der Grasnarbe bei verschiedener Nutzungsweise und Pflanzengesellschaft. Pflanzenbau *19*, 221–236 (1943)

Klebesadel, L.J.: Modification of alfalfa root system by severing the tap root. Agron. J. *56*, 359–361 (1964)

Klepper, B., Taylor, H.M., Huck, M.G., Fiscus, E.L.: Water relations and growth of cotton in drying soil. Agron. J. *65*, 307–310 (1973)

Kmoch, H.G.: Über den Umfang und einige Gesetzmäßigkeiten der Wurzelmassenbildung unter Grasnarben. Z. Acker- Pflanzenbau *95*, 363–380 (1952)

Kmoch, H.G.: Die Wurzelarbeiten J.E. Weavers und seiner Schule. Z. Acker- Pflanzenbau *104*, 275–288 (1957)

Kmoch, H.G.: Die Herstellung von Wurzelprofilen mit Hilfe des UTAH-Erdbohrers und ihre Ausdeutung. 1. Mitteilung: Methodik der Wurzelprofil-Herstellung. Z. Acker- Pflanzenbau *110*, 249–254 (1960a)

Kmoch, H.G.: Die Herstellung von Wurzelprofilen mit Hilfe des UTAH-Erdbohrers und ihre Ausdeutung. 2. Mitteilung: Zur Wurzelausbildung von Winterweizen in einer Parabraunerde. Z. Acker- Pflanzenbau *110*, 425–437 (1960b)

Kmoch, H.G.: Die Herstellung von Wurzelprofilen mit Hilfe des UTAH-Erdbohrers und ihre Ausdeutung. 3. Mitteilung: Zur Durchwurzelung verschiedener Bodentypen durch Weizen und Roggen. Z. Acker- Pflanzenbau *113*, 342–360 (1961)

Kmoch, H.G.: Die Luftdurchlässigkeit des Bodens. Ihre Bestimmung und ihre Bedeutung für einige ackerbauliche Probleme. Berlin: Gebr. Borntraeger, 1962

Kmoch, H.G.: Über Korn- und Wurzelbildung beim Winterweizen in Abhängigkeit von der Witterung und von langjährigen Düngungsmaßnahmen. Forsch. Berat. Reihe B. Ber. Landwirtsch. Fakultät Univ. Bonn, Heft *10*, 125–140 (1964a)

Kmoch, H.G.: Bodenfeuchtegang und Durchwurzelung verschiedener Bodentypen. Mitt. Dtsch. Bodenkdl. Ges. *2*, 35–45 (1964b)

Kmoch, H. G., Hanus, H.: Die Herstellung von Wurzelprofilen mit Hilfe des UTAH-Erdbohrers und ihre Ausdeutung. 4. Mitteilung: Zur Veränderlichkeit der Durchwurzelung unter dem Einfluß von Boden und Witterung. Z. Acker- Pflanzenbau *126*, 1–18 (1967)

Kmoch, H. G., Hanus, H.: Zum Einfluß der Horizontgrenzen auf die Wurzelverteilung. Z. Acker- Pflanzenbau *127*, 103–113 (1968)

Kmoch, H.G., Ramis, R.E., Fox, R.L., Koehler, F.E.: Root development of winter wheat as influenced by soil moisture and nitrogen fertilizazion. Agron. J. *49*, 20–25 (1957)

Kmoch, H.G., Halfmann, H.H., Sievers, A.: Jahreszeitliche Entwicklung der Wurzelmasse unter einer Weide in der Kölner Bucht. Z. Acker- Pflanzenbau *105*, 121–144 (1958)

Knievel, D.P.: Procedure for estimating ratio of live and dead root dry matter in root core samples. Crop Sci. *13*, 124–126 (1973)

Knop, W., Wolf, W.: Ueber Wasser- und Landwurzeln. Landwirtsch. Vers. Stn. *7*, 345–351 (1865)

Köhnlein, J.: Die Durchporung und Durchwurzelung des Unterbodens. Schriftenr. Landwirtsch. Fakultät der Christian-Albrechts-Universität Kiel, Heft *14*, 3–41 (1955)

Köhnlein, J.: Die Bedeutung der Unterbodenporung. Landwirtsch. Forsch. Sonderheft *14*, 61–71 (1960)

Köhnlein, J., Bergt, K.: Untersuchungen zur Entstehung der biogenen Durchporung im Unterboden eingedeichter Marschen. Z. Acker- Pflanzenbau *133*, 261–298 (1971)

Köhnlein, J., Vetter, H.: Ernterückstände und Wurzelbild. Hamburg-Berlin: Parey, 1953

Kokkonen, P.: Über das Verhältnis der Winterfestigkeit des Roggens zur Dehnbarkeit und Dehnungsfestigkeit seiner Wurzeln. Vorläufige Mitteilung. Acta For. Fenn. *33*, 1–45 (1927)

Kokkonen, P.: Untersuchungen über die Wurzeln der Getreidepflanzen. I. Die Wurzelformen, ihr Bau, ihre Aufgabe und Lage im Wurzelsystem. Acta For. Fenn. *37*, 1–123 (1931)

Kolek, J. (ed.): Structure and Function of Primary Root Tissues. Bratislava: Veda, Publishing House of the Slovak Academy of Sciences. 1974

Kolesnikov, V.A.: The root system of fruit tree seedlings. J. Pomol. Hortic. Sci. *8*, 197–203 (1930)

Kolesnikov, V.A.: Untersuchungen über Gesetzmäßigkeiten am Wurzelsystem bei Obstgehölzen mittels einer besonderen Methode der Probenentnahme. Tagungsber. Dtsch. Akad. Landwirtschaftswiss. Berlin *35*, 169–177 (1962a)

Kolesnikov, V.A.: Untersuchungen über das Wurzelsystem und die Ernährung der Obstgehölze. Arch. Gartenbau *10*, 447–459 (1962b)

Kolesnikov, V.A.: Neue Ergebnisse auf dem Gebiet des Wurzelwachstums im Zusammenhang mit der Bodenpflege in den Obstanlagen. Wiss. Z. Humboldt-Univ. Berlin, Math.-Nat. Reihe *18*, 735–738 (1969)

Kolesnikov, V.A.: The Root System of Fruit Plants. Moscow, USSR: Mir Publishers, 1971, 269 pp

Kolesnikov, V.A.: Methods of Studying the Root Systems of Woody Plants (Russ.). Moscow, USSR, 1972, 152 pp

Kolosov, I.I.: Absorptive Activity of Root Systems of Plants. Engl. Translation from the Russian published for the U.S. Department of Agriculture, Washington D.C. by the Indian Scientific Documentation Centre, New Dehli 1974, 525 pp

Könekamp, A.: Beitrag zur Kenntnis des Wurzelwachstums einiger Klee- und Gräserarten. Landwirtsch. Jahrb. *80*, 571–589 (1934)

Könekamp, A.: Untersuchungen über die Bewurzelung von Hülsenfrüchten. Pflanzenbau *12*, 1–11 (1936)

Könekamp, A.: Teilergebnisse von Wurzeluntersuchungen. Zeitschr. Pflanzenernähr. Düng. Bodenkde. *60*, 113–124 (1953)

Könekamp, A., Zimmer, E.: Ergebnisse der Wurzeluntersuchungen in Völkenrode 1949-1953. Zeitschr. Pflanzenernähr. Düng. Bodenkde. *68*, 158–169 (1955)

Könnecke, G.: Ertragssteigerung und Erhaltung der Bodenfruchtbarkeit durch Umstellung von Fruchtfolgen. Kühn-Archiv *64*, 107–224 (1951)

Köpke, U.: Ein Vergleich von Feldmethoden zur Bestimmung des Wurzelwachstums bei landwirtschaftlichen Kulturpflanzen. Diss. Agric. Göttingen, 1979

Köstler, J.N., Brückner, E., Bibelriether, H.: Die Wurzeln der Waldbäume. Hamburg-Berlin: Parey, 1968

Krasilnikov, P.K.: The methods of studying the subterranean parts of the trees, shrubs and forest communities in the course of the geobotanical field investigations. (Russ.). In: Polevaya Geobotanika. Lavrenko, E.M., Korchagin, A.A. (eds.). Moscow-Leningrad: Academy of Sciences of the USSR Press, 1960, Vol. 2, pp. 448–499

Kraus, C.: Untersuchungen über die Bewurzelung der Kulturpflanzen in physiologischer und kultureller Beziehung. Forsch. Geb. Agrikulturphys. *15*, 234–286 (1892), *17* 55–103 (1894) *18*, 113–166 (1895) *19*, 80–129 (1896)

Kraus, C.: Zur Kenntnis der Verbreitung der Wurzeln in Beständen von Rein- und Mischsaaten. Fühlings Landwirtsch. Zg. *63*, 337–362, 369–383, 401–412 (1914)

Krauss, G., Wobst, W., Gärtner, G.: Humusauflage und Bodendurchwurzelung im Eibenstocker Granitgebiet. Tharandter Forstl. Jahrb. *85*, 290–370 (1934)

Kreutzer, K.: Wurzelbildung junger Waldbäume auf Pseudogleyböden. Forstwiss. Centralbl. *80*, 356–392 (1961)

Kreutzer, K.: The root system of the red alder (*Alnus glutinosa* Gärtn.). In: Methods of Productivity Studies in Root Systems and Rhizosphere Organisms. Int. Symp. USSR 1968. Ed. by USSR Academy of Sciences, Leningrad: Nauka, 1968, pp. 114–119

Kroemer, K.: Beiträge zur Methodik der Wurzeluntersuchungen. Bericht der Königlichen Lehranstalt für Wein-, Obst- und Gartenbau zu Geisenheim am Rhein 1905, pp. 200–207

Kroemer, K.: Untersuchungen über das Wurzelwachstum des Weinstocks. Landwirtsch. Jahrb. *51*, 673–729 (1918a)

Kroemer, K.: Beobachtungen über die Wurzelentwicklung der Gemüsepflanzen. Landwirtsch. Jahrb. *51*, 731–745 (1918b)

Kulenkamp, A.: Methodik der Untersuchung von Apfelwurzeln. Lantbrukshögsk. Ann. *35*, 1031–1039 (1969)

Kulenkamp, A., Durmanov, D.: Geno- und phänotypische Besonderheiten der Wurzelsysteme bei immergrünen und laubabwerfenden Obstarten. Beitr. Trop. Landwirtsch. Veterinärmed. *12*, 57–68 (1974)

Kullmann, A.: Die Abhängigkeit der Bewurzelung von den Standortsbedingungen bei *Molinia caerulea*. Arch. Forstwes. *6*, 313–329 (1957a)

Kullmann, A.: Über die Wurzelentwicklung und Bestockung von *Stipa capillata* und *Molinia caerulea*. Wiss. Z. Univ. Halle, Math.-Nat. Reihe *6*, 167–176 (1957b)

Kullmann, A.: Zur Intensität der Bodendurchwurzelung. Z. Acker- Pflanzenbau *103*, 189–197 (1957c)

Kummerow, J., Krause, D., Jow, W.: Root systems of chaparral shrubs. Oecologia (Berl.) *29*, 163–177 (1977)

Kuntze, H., Neuhaus, H.: Die landeskulturelle Bedeutung der Pflanzenwurzel – ein Beitrag zum Problem der biologischen Standortverbesserung. Kulturtechniker *48*, 60–76 (1960)

Kutschera, L.: Wurzelatlas mitteleuropäischer Ackerunkräuter und Kulturpflanzen. Frankfurt/M.: DLG-Verlag, 1960

Kutschera-Mitter, L.: Über das geotrope Wachstum der Wurzel. Beitr. Biol. Pflanzen *47*, 371–436 (1971)

Kvarazkhelia, T.: Beiträge zur Biologie des Wurzelsystems der Obstbäume. Gartenbauwissenschaft *4*, 239–341 (1931)

Ladefoged, K.: Untersuchungen über die Periodizität im Ausbruch und Längenwachstum der Wurzeln bei einigen unserer gewöhnlichsten Waldbäume. Forstl. Versuchswes. Dan. *16*, 1–256 (1939)

Laird, A.S.: A study of the root systems of some important sod-forming grasses. Fla. Agric. Exp. Stn. Bull. No. 211 (1930) 28pp

Laitakari, E.: The root system of pine (*Pinus silvestris*). A morphological investigation. (Fin., Engl. Sum.). Acta For. Fenn. *33,1*, 1–380 (1929a)

Laitakari, E.: Die Wurzelforschung in ihrer Beziehung zur praktischen Forstwirtschaft. Acta For. Fenn. *33,2*, 1–21 (1929b)

Laitakari, E.: The root system of birch (*Betula verrucosa* and *odorata*). (Fin., Engl. Sum.). Acta For. Fenn. *41,2*, 1–217 (1935)

Lamba, P.S., Ahlgren, H.L., Muckenhirn, R.J.: Root growth of alfalfa, medium red clover, bromegrass, and timothy under various soil conditions. Agron. J. *41*, 451–458 (1949)

Land, W.B., Carreker, J.R.: Results of evapotranspiration and root distribution studies. Agric. Eng. *34*, 319–322 (1953)

Lang, A.R.G., Melhuish, F.M.: Lengths and diameters of plant roots in non-random populations by analysis of plane surfaces. Biometrics *26*, 421–431 (1970). *28*, 625–627 (1972)

Larson, M.M.: Construction and use of glass-faced boxes to study root development of tree seedlings. Rocky Mount. For. Range Exp. Stn. Fort Collins, Colorado, USA. Res. Note No. 73 (1962) 4 pp

Larson, M.M., Schubert, G.H.: Root competition between ponderosa pine seedlings and grass. Rocky Mount. For. Range Exp. Stn. Fort Collins, Colorado, USA. USDA Forest Service. Res. Pap. RM-54 (1969) 12 pp

Lauenroth, W.K., Whitman, W.C.: A rapid method for washing roots. J. Range Manage. *24*, 308–309 (1971)

Lavin, F.: A glass-faced planter box for field observations on roots. Agron. J. *53*, 265–268 (1961)

Lay, P.M.: Automatic sample changer for large samples. Lab. Pract. *22*, 728–730 (1973)

Ledig, F.T., Bormann, F.H., Wenger, K.F.: The distribution of dry matter growth between shoot and roots in loblolly pine. Bot. Gaz. *131*, 349–359 (1970)

Lee, H.A.: The distribution of the roots of sugar cane in the soil in the Hawaiian Islands. Plant Physiol. *1*, 363–378 (1926)

Lee, H.A., Bissinger, G.H.: The distribution of sugar cane roots in the soil on the Islands of Luzon. Sugar News *9*, 527–536 (1928)

Leibundgut, H., Dafis, S., Richard, F.: Untersuchungen über das Wurzelwachstum verschiedener Baumarten. 2. Mitteilung. Das Wurzelwachstum in einem Tonboden. Schweiz. Z. Forstwes. *114*, 621–646 (1963)

Lemke, K.: Untersuchungen über das Wurzelsystem der Roteiche auf diluvialen Standortsformen. Arch. Forstwes. *5*, 8–45, 161–202 (1956)

Leonard, O.A.: Cotton root development in relation to natural aeration of some Mississippi blackbelt and delta soils. J. Am. Soc. Agron. *37*, 55–71 (1945)

Lesczynski, D.B., Tanner, C.B.: Seasonal variation of root distribution of irrigated, field-grown Russet Burbank potato. Am. Potat. J. *53*, 69–78 (1976)

Levin, I., Bravdo, B., Assaf, R.: Relation between apple root distribution and soil water extraction in different irrigation regimes. In: Physical Aspects of Soil Water and Salts in Ecosystems. Hadas, A., Swartzendruber, D., Rijtema, P.E., Fuchs, M., Yaron, B. (eds.). (Ecological Studies *4*). Berlin-Heidelberg-New York: Springer, 1973, pp. 351–359

Lichtenegger, E.: Wurzelbild und Lebensraum. Beitr. Biol. Pflanz. *52*, 31–56 (1976)

Lieshout, van, J.W.: Root development of some field crops. (Dutch, Engl. Sum.). Versl. Landbouwk. Onderz. *62,16*, (1956) 46 pp

Lieshout, van, J.W.: Effect of soil conditions on development and activity of the root system. (Dutch, Engl. Sum.). Versl. Landbouwk. Onderz. *66,18* (1960) 91 pp

Lieth, H.: The determination of plant dry-matter production with special emphasis on the underground parts. In: Functioning of Terrestrial Ecosystems at the Primary Production Level. Proceedings of the Copenhagen Symposium. Eckardt, F.E. (ed.). Paris: UNESCO 1968, pp. 179–186

Lieth, H., Whittaker, R.H. (ed.): Primary Productivity of the Biosphere. (Ecological Studies Vol. 14). Berlin-Heidelberg-New York: Springer, 1975

Linford, M.B., Rhoades, H.L.: Centrifugation of roots before determining moist weight. Plant Dis. Rep. *43*, 987–988 (1959)

Linkola, K., Tiirikka, A.: Über Wurzelsysteme und Wurzelausbreitung der Wiesenpflanzen auf verschiedenen Standorten. (Germ., Finn. Sum.). Ann. Bot. Soc. Zool. Bot. Fenn. Vanamo. Vol. *6*, No. 6 (1936) 207 pp

Linscott, D.L., Fox, R.L., Lipps, R.C.: Corn root distribution and moisture extraction in relation to nitrogen fertilization and soil properties. Agron. J. *54*, 185–189 (1962)

Lippert, G.: Beitrag zur Wurzelentwicklung der Gräser in Reinbeständen unter Einwirkung der Beregnung und der Stickstoffdüngung. Kulturtechniker *47*, 251–262 (1959)

Lipps, R.C., Fox, R.L.: Root activity of sub-irrigated alfalfa as related to soil moisture, temperature, and oxygen supply. Soil Sci. *97*, 4–12 (1964)

Lipps, R.C., Fox, R.L., Koehler, F.E.: Characterizing root activity of alfalfa by radioactive tracer techniques. Soil Sci. *84*, 195–204 (1957)
Litav, M., Harper, J.L.: A method for studying spatial relationships between the root systems of two neighbouring plants. Plant Soil *26*, 389–392 (1967)
Livingston, B.E.: Note on the relation between the growth of roots and of tops in wheat. Bot. Gaz. *31*, 139–143 (1906)
Long, O.H.: Root studies on some farm crops in Tennessee. University of Tennessee. Agric. Exp. Stn. Bull. No. 301 (1959) 41 pp
Longenecker, D., Merkle, F.G.: Influence of placement of lime compounds on root development and soil characteristics. Soil Sci. *73*, 71–74 (1952)
Loeters, J.W.J., Bakermans, W.A.P.: De invloed van enkele groenbemestingsgewassen en hun beworteling op de structuur van zandgronden. Meded. Dir. Tuinbouw *27*, 565–572 (1964)
Loeters, J.W.J., Bakermans, W.A.P., van der Zweerde, H.: Invloed van diepe grondbewerking op bewortelbaarheid van een zandondergrond. Landbouwvoorlichting *26*, 360–368 (1969)
Lott, W.L., Satchell, D.P., Hall, N.S.: A tracer-element technique in the study of root extension. Proc. Am. Soc. Hortic. Sci. *55*, 27–34 (1950)
Ludwig, J.A.: Distributional adaptations of root systems in desert environments. In: The Belowground Ecosystem: A Synthesis of Plant-Associated Processes. Marshall, J.K. (ed.). Range Science Department Science Series No. 26, Colorado State University, Fort Collins, USA, 1977, pp. 85–91
Lund, Z.F., Beals, H.O.: A technique for making thin sections of soil with roots in place. Proc. Soil Sci. Soc. Am. *29*, 633–635 (1965)
Lund, Z.F., Pearson, R.W., Buchanan, G.A.: An implanted soil mass technique to study herbicide effects on root growth. Weed Sci. *18*, 279–281 (1970)
Lungley, D.R.: The growth of root systems—a numerical computer simulation model. Plant Soil *38*, 145–159 (1973)
Lupton, F.G.H., Oliver, R.H., Ellis, F.B., Barnes, B.T., Howse, K.R., Welbank, P.J., Taylor, P.J.: Root and shoot growth of semi-dwarf and taller winter wheats. Ann. Appl. Biol. *77*, 129–144 (1974)
Lutz, H.J., Ely, J.B., Little, S.: The influence of soil profile horizons on root distribution of white pine (*Pinus strobus* L.). New Haven: Yale University – Sch. For. Bull. No. 44 (1937)
Lyford, W.H.: Rhizography of non-woody roots of trees in the forest floor. In: The Development and Function of Roots. Torrey, J.G., Clarkson, D.T. (eds.). London-New York-San Francisco: Academic Press, 1975, pp. 179–196
Lyford, W.H., Wilson, B.F.: Development of the root system of *Acer rubrum* L. Harvard University Petersham, Mass., USA. Harvard For. Pap. No. *10* (1964) 17 pp
Lyford, W., Wilson, B.F.: Controlled growth of forest tree roots: Technique and application. Harvard University Petersham, Mass., USA. Harvard For. Pap. No. *16* (1966) 12 pp
Lyons, C.G., Krezdorn, A.H.: Distribution of peach roots in lakeland fine sand and the influence of fertility levels. Proc. Fla. State Hortic. Soc. *75*, 371–377 (1962)
Lyr, H., Hoffmann, G.: Untersuchungen über das Wurzel- und Sprosswachstum einiger Gehölze. Silva Fenn. *117* (4), 1–19 (1965)
Lyr, H., Hoffmann, G.: Growth rates and growth periodicity of tree roots. Int. Rev. For. Res. *2*, 181–236 (1967)
Maas, G.: Über Schäden an den Ankerwurzeln (Kronenwurzeln) des Getreides nach Herbizid-Behandlung. Z. Pflanzenkr. Pflanzenschutz, Sonderheft *4*, 139–144 (1968)
Maas, G.: Über den Einfluß von Herbiziden auf die Standfestigkeit von Getreide. Z. Pflanzenkr. Pflanzenschutz, Sonderheft *5*, 129–135 (1970)
Mac Key, J.: The wheat root. Proc. 4th Int. Wheat Genet. Symp. Missouri Agric. Exp. Sta., Columbia, Mo. (1973) pp. 827–842
Magnaye, A.B.: Studies on the root system of healthy and cadang-cadang affected coconut trees. Philippine J. Plant Ind. *34*, 143–151 (1969)
Malicki, L.: Die Wurzelmengenbestimmung in Feldbedingungen. (Pol., Germ. Sum.). Zesz. Probl. Postepów Nauk Roln. *88*, 17–31 (1968)

Markle, M.S.: Root systems of certain desert plants. Bot. Gaz. *64*, 177–205 (1917)

Marks, G.C., Kozlowski, T.T. (eds.): Ectomycorrhizae. Their Ecology and Physiology. New York, London: Academic Press, 1973

Marsh, B.a'B.: Measurement of length in random arrangements of lines. J. Appl. Ecol. *8*, 265–267 (1971)

Marshall, J.K. (ed.): The Belowground Ecosystem: A Synthesis of Plant-Associated Processes. Range Science Department Science Series No. 26, Colorado State University, Fort Collins, USA, 1977, 351 pp

Martin, G.C., Locascio, S., Langston, R.: An inexpensive root temperature control apparatus. Proc. Am. Soc. Hortic. Sci. *81*, 562–564 (1962)

Martin, N.E., Hendrix, J.W.: An apparatus for the mist culture of wheat. Plant Dis. Rep. *50*, 369–371 (1966)

Maschhaupt, J.G.: De beworteling onzer cultuurgewassen. Versl. Landbouwk. Onderz. *16*, 76–89 (1915)

Mathers, A.C.: Effect of radial restriction on lateral growth of the root-shoot axis of young cotton plants. Agron. J. *59*, 379–381 (1967)

Maurya, P.R., Ghildyal, B.P., Sharma, D.: Note on the determination of specific activity of ^{32}P for the study of root distribution in soil-root cores. Indian J. Agric. Sci. *43*, 886–887 (1973)

Mayaki, W.C., Teare, I.D., Stone, L.R.: Top and root growth of irrigated and nonirrigated soybeans. Crop Sci. *16*, 92–94 (1976)

McClure, J.W., Harvey, C.: Use of radiophosphorus in measuring root growth of sorghums. Agron. J. *54*, 457–459 (1962)

McDougall, W.B.: The growth of forest tree roots. Am. J. Bot. *3*, 384–392 (1916)

McElgunn, J.D., Harrison, C.M.: Formation, elongation, and longevity of barley root hairs. Agron. J. *61*, 79–81 (1969)

McElgunn, J.D., Lawrence, T.: A minirhizotron for in situ observation of seedling development. Can. J. Plant Sci. *50*, 754–756 (1970)

McKell, C.M.: Root studies. In: Pasture and Range Research Techniques. Prepared by a Joint Committee. Ithaca, New York: Comstock Publishing Associates (Cornell University Press) 1962, pp. 173–179

McKell, C.M., Wilson, A.M., Jones, M.B.: A flotation method for easy separation of roots from soil samples. Agron. J. *53*, 56–57 (1961)

McKell, C.M., Jones, M.B., Perrier, E.R.: Root production and accumulation of root material on fertilized annual range. Agron. J. *54*, 459–462 (1962)

McMinn, R.G.: Characteristics of Douglas-fir root systems. Can. J. Bot. *41*, 105–122 (1963)

McQueen, D.R.: The quantitative distribution of absorbing roots of *Pinus silvestris* and *Fagus sylvatica* in a forest succession. Oecol. Plant. *3*, 83–99 (1968)

Mederski, H.J., Wilson, J.H.: Relation of soil moisture to ion absorption by corn plants. Proc. Soil Sci. Soc. Am. *24*, 149–152 (1960)

Melhuish, F.M.: A precise technique for measurement of roots and root distribution in soils. Ann. Bot. N.S. *32*, 15–22 (1968)

Melhuish, F.M., Lang, A.R.G.: Quantitative studies of roots in soil. I. Length and diameter of cotton roots in a clay-loam soil by analysis of surface-ground blocks of resin-impregnated soil. Soil Sci. *106*, 16–22 (1968)

Melhuish, F.M., Lang, A.R.G.: Quantitative studies of roots in soil. II. Analysis of non-random populations. Soil Sci. *112*, 161–166 (1971)

Melzer, E.W.: Beiträge zur Wurzelforschung. Untersuchungen auf meliorierten und nichtmeliorierten Standorten der Oberförsterei Adorf/Vogtland. Diss. Silv. Techn. Univ. Dresden. Maschinenschriftl. vervielfältigt (1962a)

Melzer, E.W.: Die stochastischen Beziehungen zwischen Sproß- und Wurzelsystem des Baumes. Arch. Forstwes. *11*, 822–838 (1962b)

Melzer, E.W.: Die Zwischenflächendurchwurzelung in Nadelholzbeständen der Oberförsterei Adorf (Vogtland). Wiss. Z. Techn. Univ. Dresden *12*, 1569–1577 (1963)

Mengel, D.B., Barber, S.A.: Development and distribution of the corn root system under field conditions. Agron. J. *66*, 341–344 (1974a)

Mengel, D.B., Barber, S.A.: Rate of nutrient uptake per unit of corn root under field conditions. Agron. J. *66*, 399–402 (1974b)
Meredith, H.L., Patrick, W.H.: Effects of soil compaction on subsoil root penetration and physical properties of three soils in Louisiana. Agron. J. *53*, 163– 167 (1961)
Metsävainio, K.: Untersuchungen über das Wurzelsystem der Moorpflanzen. (Germ., Fin. Sum.). Ann. Bot. Soc. Zool. Bot. Fenn. Vanamo. Vol. *1*, No. 1, 1931 (422 pp)
Meusel, H.: Bewurzelung und Wuchsform einiger *Asteriscus*-Arten. Flora *150*, 441–453 (1961)
Meyer, F.H.: Feinwurzelverteilung bei Waldbäumen in Abhängigkeit vom Substrat. Forstarchiv *38*, 286–290 (1967)
Meyer, F.H., Göttsche, D.: Distribution of root tips and tender roots of beech. In: Integrated Experimental Ecology. Ellenberg, H. (ed.). (Ecological Studies *2*). Berlin-Heidelberg-New York: Springer, 1971, pp. 48–52
Michael, G., Bergmann, W.: Bodenkohlensäure und Wurzelwachstum. Z. Pflanzenernähr. Düng. Bodenkde. *65*, 180–194 (1954)
Miller, E.C.: Comparative study of the root systems and leaf areas of corn and the sorghums. J. Agric. Res. *6*, 311–332 (1916)
Miller, M.H., Mamaril, C.P., Blair, G.J.: Ammonium effects on phosphorus absorption through pH changes and phosphorus precipitation at the soil-root interface. Agron. J. *62*, 524–527 (1970)
Mishustin, E.N., Shil'nikova, V.K.: Biological Fixation of Atmospheric Nitrogen. London: Macmillan, 1971
Misra, D.K.: Relation of root development to drought resistance of plants. Indian J. Agron. *1*, 41–46 (1956)
Mitchell, R.L., Russel, W.J.: Root development and rooting patterns of soybean [*Glycine max*. (L.) Merrill] evaluated under field conditions. Agron. J.*63*, 313–316 (1971)
Mitchell, W.H.: Influence of nitrogen and irrigation on the root and top growth of forage crops. University of Delaware. Agric. Exp. Stn. Techn. Bull. No. 341 (1962) 32 pp
Mohr, D.: Untersuchungen der Wurzelverteilung von Kulturpflanzen in verschiedenen Böden mit Hilfe von radioaktivem Phosphor. Diss. Nat. Gießen (1975)
Moir, W.H., Bachelard, E.P.: Distribution of fine roots in three *Pinus radiata* plantations near Canberra, Australia. Ecology *50*, 658–662 (1969)
Molz, F.J.: Interaction of water uptake and root distribution. Agron. J. *63*, 608–610 (1971)
Monyo, J.H., Whittington, W.J.: Genetic analysis of root growth in wheat. J. Agric. Sci. *74*, 329–338 (1970)
Moraghan, J.T.: Excavation and permanent mounting of corn root systems. Agron. J. *60*, 581–582 (1968)
Morrison, J.K., Armson, K.A.: The rhizometer – a device for measuring roots of tree seedlings. For. Chron. *44*, 21–23 (1968)
Morrow, R.R.: Periodicity and growth of sugar maple surface layer roots. J. For. *48*, 875–881 (1950)
Mosher, P.N., Miller, M.H.: Influence of soil temperature on the geotropic response of corn roots (*Zea mays* L.). Agron. J. *64*, 459–462 (1972)
Mosse, B.: Advances in the study of vesicular-arbuscular mycorrhiza. Ann. Rev. Phytopathol. *11*, 171–196 (1973)
Muller, C.H.: Root development and ecological relations of guayule. US Department of Agriculture. Techn. Bull. No. 923 (1946) 114 pp
Murdoch, C.L., Criley, R.A., Fukuda, S.K.: An inexpensive method for the study of root growth in containers of solid media. Hortscience *9*, 221 (1974)
Musick, G.J., Fairchild, M.L., Fergason, V.L., Zuber, M.S.: A method of measuring root volume in corn (*Zea mays* L.). Crop. Sci. *5*, 601–602 (1965)
Muzik, T.J., Whitworth, J.W.: A technique for the periodic observation of root systems in situ. Agron. J. *54*, 56 (1962)
Najmr, S.: Über die Methode der Isolierung der Wurzelsysteme der Futterpflanzen aus dem Boden. Z. Acker- Pflanzenbau *104*, 103–109 (1957)
Nakayama, F.S., van Bavel, C.H.M.: Root activity distribution patterns of sorghum and soil moisture conditions. Agron. J. *55*, 271–274 (1963)

Nass, H.G., Zuber, M.S.: Correlation of corn (*Zea mays* L.) roots early in development to mature root development. Crop Sci. *11*, 655–658 (1971)

Neilson, J.A.: Autoradiography for studying individual root systems in mixed herbaceous stands. Ecology *45*, 644–646 (1964)

Nelson, W.W., Allmaras, R.R.: An improved monolith method for excavating and describing roots. Agron. J. *61*, 751–754 (1969)

Newbould, P.J.: Methods of estimating root production. In: Functioning of Terrestrial Ecosystems at the Primary Production Level. Proceedings of the Copenhagen Symposium. Eckardt, F.E. (ed.). Paris: UNESCO, 1968, pp. 187–190

Newman, E.I.: A method of estimating the total length of root in a sample. J. Appl. Ecol. *3*, 139–145 (1966a)

Newman, E.I.: Relationship between root growth of flax (*Linum usitatissimum*) and soil water potential. New Phytol. *65*, 273–283 (1966b)

Nilsson, H.E.: Preliminary report on a method for studies of root development and root diseases. Phytopathol. Z. *53*, 190–194 (1965)

Nilsson, H.E.: Studies of root and foot rot diseases of cereals and grasses. I. On resistance to *Ophiobolus graminis* Sacc. Lantbrukshögsk. Ann. *35*, 275–807 (1969)

Nilsson, H.E.: Method for the study of roots and root diseases under controlled culture conditions. Swed. J. Agric. Res. *3*, 79–88 (1973)

Nobbe, F.: Ueber die feinere Verästelung der Pflanzenwurzel. Landwirtsch. Vers. Stn. *4*, 212–224 (1862)

Nobbe, F.: Ueber die Pflanzenzucht im Wasser und ihre praktische Bedeutung für die Landwirthschaft. Chem. Ackersmann *15*, 65–81 (1869)

Nobbe, F.: Beobachtungen und Versuche über die Wurzelbildung der Nadelhölzer. Tharandter Forstl. Jahrb. *25*, 201–218 (1875)

Novoselov, V.S.: A closed volumeter for plant root system. (Soviet) Plant Physiol. (Engl. translation from Fiziologiya Rastenii) *7*, 203–204 (1960)

Nutman, F.J.: The root-system of *Coffea arabica*. I. Root-systems in typical soils of British East Africa. Emp. J. Exp. Agric. *1*, 271–284 (1933a)

Nutman, F.J.: The root-system of *Coffea arabica*. II. The effect of some soil conditions in modifying the 'normal' root-system. Emp. J. Exp. Agric. *1*, 285–296 (1933b)

Nutman, F.J.: The root-system of *Coffea arabica*. III. The spatial distribution of the absorbing area of the root. Emp. J. Exp. Agric. *2*, 293–302 (1934)

Nye, P.H., Tinker, P.B.: The concept of a root demand coefficient. J. Appl. Ecol. *6*, 293–300 (1969)

Oberländer, H.-E., Zeller, A.: Die Phosphataufnahme durch verschiedene Zonen des Wurzelsystems der Luzerne. Bodenkultur *15*, 317–328 (1964)

Obermayer, E.: Root study technique. Herb. Rev. *7*, 175 (1939)

Olivin, J.: An observation method of the oil palm root system. (Fr., Engl. Sum.) Oléagineux *20*, 731–733 (1965)

Olson, J.S.: Distribution and radiocesium transfers of roots in a tagged mesophytic Appalachian forest in Tennessee. In: Methods of Productivity Studies in Root Systems and Rhizosphere Organisms. Int. Symp. USSR, 1968. Ed. by USSR Academy of Sciences, Leningrad: Nauka, 1968, pp. 133–139

Onderdonk, J.J., Ketcheson, J.W.: Effect of soil temperature on direction of corn root growth. Plant Soil *39*, 177–186 (1973a)

Onderdonk, J.J., Ketcheson, J.W.: Effect of stover mulch on soil temperature, corn root weight, and phosphorus fertilizer uptake. Proc. Soil Sci. Soc. Am. *37*, 904–906 (1973b)

Opitz, K.: Untersuchungen über Bewurzelung und Bestockung einiger Getreidesorten. Mitt. Landwirtsch. Inst. Univ. Breslau *2*, 749–816 (1904)

Opitz von Boberfeld, W.: Zur Problematik des Stichprobenumfanges bei Wurzelgewichtsfeststellungen von Rasengräsern. Rasen-Turf-Gazon *3*, 51–53 (1972)

Opitz von Boberfeld, W., Boeker, P.: Der Einfluß verschiedener Düngemittel auf die Anhäufung der Wurzelmasse eines Intensivrasentyps. Rasen-Turf-Gazon *4*, 25–27 (1973)

Ortman, E.E., Peters, D.C., Fitzgerald, P.J.: Vertical-pull technique for evaluating tolerance of corn root systems to Northern and Western corn rootworms. J. Econ. Entomol. *61*, 373–375 (1968)

Oskamp, J.: The rooting habit of deciduous fruits on different soils. Proc. Am. Soc. Hortic. Sci. *29*, 213–219 (1932)

Oskamp, J.: Root studies of young apple trees. Gartenbauwissenschaft *7*, 7–14 (1933)

Oskamp, J.: Soils in relation to fruit growing in New York. Part V. The vineyard soils of the Westfield Area, Chautauqua County. Cornell Univ. Agric. Exp. Stn. Ithaca, New York. Bull. 609 (1934) 18 pp

Oskamp, J.: Soils in relation to fruit growing in New York. Part VI. Tree behavior on important soil profiles in the Williamson-Marion Area, Wayne County. Cornell Univ. Agric. Exp. Stn. Ithaca, New York. Bull. 626 (1935a) 29 pp

Oskamp, J.: Soils in relation to fruit growing in New York. Part VIII. Tree behavior on important soil profiles in the Medina-Lyndonville-Carlton Area, Orleans County. Cornell Univ. Agric. Exp. Stn. Ithaca, New York. Bull. 633 (1935b) 21 pp

Oskamp, J.: Soils in relation to fruit growing in New York. Part IX. Tree behavior on important soil profiles in the Newfane-Olcott Area, Niagara County. Cornell Univ. Agric. Exp. Stn. Ithaca, New York. Bull. 653 (1936) 20 pp

Oskamp, J., Batjer, L.P.: Soils in relation to fruit growing in New York. Part II. Size, production, and rooting habit of apple trees on different soil types in the Hilton and Morton Areas, Monroe County. Cornell Univ. Agric. Exp. Stn. Ithaca, New York. Bull. 550 (1932) 45 pp

Oskamp, J., Batjer, L.P.: Soils in relation to fruit growing in New York. Part III. Some physical and chemical properties of the soils in the Hilton and Morton Areas, Monroe County, and their relation to orchard performance. Cornell Univ. Agric. Exp. Stn. Ithaca, New York. Bull. 575 (1933) 34 pp

Osman, A.Z.: Der jahreszeitliche Verlauf der Wurzelaktivität von *Dactylis glomerata* L. und *Ranunculus frieseanus* Jord. in Rein- und Mischkultur, gemessen mit ^{32}P. Schweiz. Landwirtsch. Forsch. *11*, 159–189 (1972)

Ostermann, W.: Vergleichende morphologische und physiologische Untersuchungen am Wurzelsystem verschiedener Kartoffelsorten. Angew. Bot. *13*, 297–337 (1931)

Ostermayer, A.: Pflanzenzüchtung auf Widerstandsfähigkeit gegen Trockenperioden. Züchter *6*, 155–162 (1934)

Otto, G.: Ein Quotient zur Bestimmung der Verzweigungsdichte von Gehölzfaserwurzeln im Zusammenhang mit Problemen der Rhizosphäre. Deutsche Akademie der Landwirtschaftswissenschaften zu Berlin. Tagungsberichte Nr. 35, 179–184 (1962)

Otto, G.: Kennzahlen des Faserwurzelwachstums, eine Voraussetzung für die Prüfung des Einflusses von Mikroorganismen auf das Wurzelwachstum. Zentralbl. Bakteriol. Parasitenkd. Infektionskr. Hyg. II. Abt. *116*, 512–516 (1963)

Otto, G.: Der Einfluss verschiedener Bodenfaktoren auf die Verzweigungsdichte von Gehölzfaserwurzeln. Deutsche Akademie der Landwirtschaftswissenschaften zu Berlin. Tagungsberichte Nr. 65, 259–264 (1964)

Ovington, J.D., Murray, G.: Seasonal periodicity of root growth of birch trees. In: Methods of Productivity Studies in Root Systems and Rhizosphere Organisms. Int. Symp. USSR, 1968. Ed. by USSR Academy of Sciences, Leningrad: Nauka, 1968, pp. 146–154

Ozanne, P.G., Asher, C.J., Kirton, D.J.: Root distribution in a deep sand and its relationship to the uptake of added potassium by pasture plants. Aust. J. Agric. Res. *16*, 785–800 (1965)

Paasikallio, A.: Radiophosphorus used to follow root growth in field plantings of spring cereals. Ann. Agric. Fenn. *15*, 175–181 (1976)

Paavilainen, E.: On the effect of drainage on root systems of Scots pine on peat soils. (Fin., Engl. Sum.). Commun. Inst. For. Fenn. *61,1*, (1966) 110 pp

Page, E.R., Gerwitz, A.: Mathematical models, based on diffusion equations, to describe root systems of isolated plants, row crops, and swards. Plant Soil *41*, 243–254 (1974)

Parao, F.T., Paningbatan, E. Jr., Yoshida, S.: Drought resistance of rice varieties in relation to their root growth. Philippine J. Crop Sci. *1*, 50–55 (1976)

Parker, J.: Effects of variations in the root-leaf ratio on transpiration rate. Plant Physiol. *24*, 739–743 (1949)

Parmeijer, W.L.: Painting exposed roots in situ for photographing. Queensl. J. Agric. Sci. *13*, 66–67 (1956)

Parsche, F.: Wurzelraum- und Nährstoffkontrolle in Trockengebieten. Forschungsdienst *12*, 61–74 (1941)

Partridge, N.L.: A container for growing plants for root studies. J. Am. Soc. Agron. *32*, 907–908 (1940)

Patel, G.C.: Root distribution and moisture extraction pattern of flue-cured tobacco under dry conditions. Indian Tobacco *14*, 187–197 (1964)

Pätzold, H.: Der Stand der Wurzelforschung bei Futterpflanzen. Dtsch. Landwirtsch. *14*, 497–499 (1963a)

Pätzold, H.: Die Anwendung der Ergebnisse der Wurzelforschung für den Ackerbau. Dtsch. Landwirtsch. *14*, 581–584 (1963b)

Pavlychenko, T.K.: The soil-block washing method in quantitative root study. Can. J. Res. Sect. C, *15*, 34–57 (1937a)

Pavlychenko, T.K.: Quantitative study of the entire root systems of weed and crop plants under field conditions. Ecology *18*, 62–79 (1937b)

Pavlychenko, T.: Root systems of certain forage crops in relation to the management of agricultural soils. Nat. Res. Counc. Can. Dominion Dep. Agric. Ottawa. N.R.C. No. 1088 (1942) 46pp

Pearson, R.W.: Soil environment and root development. In: Plant Environment and Efficient Water Use. Ed. by Pierre, W.H., Kirkham, D., Pesek, J., Shaw, R. (eds.). Madison, Wisc., USA: Am. Soc. Agron. Soil Sci. Soc. Am. 1966, pp. 95–126

Pearson, R.W.: Significance of rooting pattern to crop production and some problems of root research. In: The Plant Root and its Environment. Carson, E.W. (ed.). Charlottesville, USA: University Press of Virginia, 1974, pp. 247–270

Pearson, R.W., Lund, Z.F.: Direct observation of cotton root growth under field conditions. Agron. J. *60*, 442–443 (1968)

Pearson, R.W., Childs, J., Lund, Z.F.: Uniformity of limestone mixing in acid subsoil as a factor in cotton root penetration. Proc. Soil. Sci. Soc. Am. *37*, 727–732 (1973)

Petrov, M.P.: The results of investigation in root systems of desert plants in Middle Asia. In: Methods of Productivity Studies in Root Systems and Rhizosphere Organisms. Int. Symp. 1968. Ed. by USSR Academy of Sciences, Leningrad: Nauka, 1968, pp. 158–161

Pettit, R.D., Jaynes, C.C.: Use of radiophosphorus and soil-block techniques to measure root development. J. Range Manage. *24*, 63–65 (1971)

Phillips, J.M., Hayman, D.S.: Improved procedures for clearing roots and staining parasitic and vesicular-arbuscular mycorrhizal fungi for rapid assessment of infection. Trans. Brit. Mycolog. Soc. *55*, 158–161 (1970)

Pickett, S.T.A.: Distribution and interactions of surface roots of *Castilla elastica* (Moraceae) in lowland Costa Rica. Turrialba *26*, 156–159 (1976)

Pinkas, L.L.H., Teel, M.R., Swartzendruber, D.: A method of measuring the volume of small root systems. Agron. J. *56*, 90–91 (1964)

Pinthus, M.J.: Spread of the root system as indicator for evaluating lodging resistance of wheat. Crop Sci. *7*, 107–110 (1967)

Pinthus, M.J., Eshel, Y.: Observations on the development of the root system of some wheat varieties. Isreal J. Agric. Res. *12*, 13–20 (1962)

Pistohlkors, von, H.: Wurzelkenntnis und Pflanzenproduktion. Bonn: O. Paul und Riga: N. Kymmel, 1898

Pittman, U.J.: Growth reaction and magnetotropism in roots of winter wheat (Kharkov 22 M.C.). Can. J. Plant Sci. *42*, 430–436 (1962)

Polle, R.: Über den Einfluß verschieden hohen Wassergehalts, verschiedener Düngung und Festigkeit des Bodens auf die Wurzelentwicklung des Weizens und der Gerste im ersten Vegetationsstadium. J. Landwirtsch. *58*, 297–344 (1910)

Portas, C.A.M.: Development of root systems during the growth of some vegetable crops. Plant Soil *39*, 507–518 (1973)

Post, van der, C.J.: Simultaneous observations on root and top growth. Acta Hortic. Techn. Commun. No. *7*, 138–145 (1968)

Post, van der, C.J., Groenewegen, C.L.: Root development of lettuce. (Dutch, Engl. Sum.). Meded. Dir. Tuinbouw *23*, 107–117 (1960)

Post, van der, C.J., Meijs, van der, M.Q.: Relationships between root growth and crop development of some vegetables under glass. (Dutch, Engl. Sum.). Meded. Dir. Tuinbouw *31*, 447–453 (1968)

Post, van der, C.J., Meijs, van der, M.Q.: Methods of investigating root growth on vegetables grown under glass. In: Root Growth. Whittington, W.J. (ed.). London: Butterworth, 1969, pp. 404–405

Preston, R.J.: The growth and development of root systems of juvenile lodgepole pine. Ecol. Monographs. *12*, 449–468 (1942)

Price, K.R.: A field method for studying root systems. Health Phys. *11*, 1521–1525 (1965)

Priestley, J.H., Pearsall. W.H.: Growth studies. III. A volumeter method of measuring the growth of roots. Ann. Bot. *36*, 485–488 (1922)

Przemeck, E., Alcalde-Blanco, S.: Ein Hydro-Sprühkulturverfahren zum Studium ernährungsphysiologischer Probleme. Angew. Bot. *43*, 331–339 (1969)

Racz, G.J., Rennie, D.A., Hutcheon, W.L.: The P 32 injection method for studying the root system of wheat. Can. J. Soil Sci. *44*, 100–108 (1964)

Raper, C.D., Jr., Barber, S.A.: Rooting systems of soybeans. I. Differences in root morphology among varieties. Agron. J. *62*, 581–584 (1970a)

Raper, C.D., Jr., Barber, S.A.: Rooting system of soybeans. II. Physiological effectiveness as nutrient absorption surfaces. Agron. J. *62*, 585–588 (1970b)

Rauhe,, K.: Einfluß der Tiefkultur auf die Wurzelentwicklung einiger Kulturpflanzen der leichten Böden. Deutsche Akademie der Landwirtschaftswissenschaften zu Berlin. Wiss. Abh. Nr. 51, 42–60 (1961)

Rauhe, K.: Untersuchungen über den Einfluß bestimmter Tiefkulturmaßnahmen auf leichten Böden mit besonderer Berücksichtigung des Pflanzenertrages, der physikalischen Bodeneigenschaften und der Wurzelentwicklung. Wiss. Z. Humboldt-Univ. Berlin, Math.-Nat. Reihe *11*, 99–134 (1962)

Reicosky, D.C., Millington, R.J., Peters, D.B.: A comparison of three methods for estimating root length. Agron. J. *62*, 451–453 (1970)

Reicosky, D.C., Millington, R.J., Klute, A., Peters, D.B.: Patterns of water uptake and root distribution of soybeans *(Glycine max.)* in the presence of a water table. Agron. J. *64*, 292–297 (1972)

Reid, D.A., Fleming, A.L., Foy, C.D.: A method of determining aluminium response of barley in nutrient solution in comparison to response in Al-toxic soil. Agron. J. *63*, 600–603 (1971)

Reijmerink, A.: A new method for recording root distribution. (Dutch, Engl. Sum.) Meded. Dir. Tuinbouw *27*, 42–49 (1964)

Reijmerink, A.: Microstructure, soil strength and root development of asparagus on loamy sands in the Netherlands. Neth. J. Agric. Sci. *21*, 24–43 (1973)

Reitz, H.J., Long, W.T.: Water table fluctuation and depth of rooting of citrus trees in the Indian river area. Proc. Fl. State Hortic. Soc. *68*, 24–29 (1955)

Reynolds, E.R.C.: Root distribution and the cause of its spatial variability in *Pseudotsuga taxifolia* (Poir.). Britt. Plant Soil *32*, 501–517 (1970)

Reynolds, E.R.C.: The distribution pattern of fine roots of trees. In: Ökologie und Physiologie des Wurzelwachstums. II. Int. Symp. Hoffmann, G. (ed.). Berlin: Akademie-Verlag 1974, pp. 101–112

Reynolds, E.R.C.: Tree rootlets and their distribution. In: The Development and Function of Roots. Torrey, J.G., Clarkson, D.T. (eds.). London-New York-San Francisco: Academic Press, 1975, pp. 163–177

Richards, D., Cockroft, B.: Soil physical properties and root concentration in an irrigated peach orchard. Aust. J. Exp. Agric. Anim. Husb. *14*, 103–107 (1974)

Richardson, S.D.: Root growth of *Acer pseudoplatanus* L. in relation to grass cover and nitrogen deficiency. Meded. Landbouwhogesch. Wageningen, Nederland *53* (4), 75–97 (1953)

Riedacker, A.: Wurzel- und Sproßwachstum von *Eucalyptus camadulensis* Dehn. im Mittelmeergebiet. In: Ökologie und Physiologie des Wurzelwachstums. II. Int. Symp. Hoffmann, G. (ed.). Berlin: Akademie-Verlag, 1974, pp. 283–296

Rieley, J.O., Summerfield, R.J.: The use of inert polymers in hydroponic studies. Plant Soil *37*, 183–185 (1972)

Rios, M.A., Pearson, R.W.: The effect of some chemical environmental factors on cotton root behavior. Proc. Soil. Sci. Soc. Am. *28*, 232–235 (1964)

Rivers, G.W., Faubion, J.L.: A uniform technique for comparing root systems of sesame *Sesamum indicum* L. Crop. Sci. *3*, 182–183 (1963)

Roach, W.A.: Plant injection as a physiological method. Ann. Bot. N.S. *3*, 155–226 (1939)

Roach, W.A., Neve, R., Vanstone, F.H., Philcox, H.J., Delap, A.V., Ford, E.M.: A method of growing apple trees by spraying their roots with nutrient solution. J. Hortic. Sci. *32*, 85–97 (1957)

Roberts, J.: A study of root distribution and growth of *Pinus sylvestris* L. (Scots pine) plantation in East Anglia. Plant Soil *44*, 607–621 (1976)

Roberts, R.H., Struckmeyer, B.E.: The effect of top environment and flowering upon top-root ratios. Plant Physiol. *21*, 332–344 (1946)

Robertson, J.H.: Penetration of roots of tall wheatgrass in wet salin-alkali soil. Ecology *36*, 755–757 (1955)

Roder, W.: Methodik und Ergebnisse der Prüfung züchterisch wertvoller Einzelpflanzennachkommenschaften von Schafschwingel, *Festuca ovina* L., auf ihre Wurzelleistung bei Drillsaat. Z. Landwirtsch. Vers. Untersuchungswes. *5*, 122–131 (1959)

Roemer, Th.: Der Einfluß der Phosphorsäure auf die Entwicklung der Wurzeln. Superphosphat *8*, 73–75 (1932)

Rogers, W.S.: The root development of an apple tree in a wet clay soil. J. Pomol. Hortic. Sci. *10*, 219–227 (1932)

Rogers, W.S.: Root studies. III. Pear, gooseberry and black currant root systems under different soil fertility conditions, with some observations on root stock and scion effect in pears. J. Pomol. Hortic. Sci. *11*, 1–18 (1933)

Rogers, W.S.: Root studies. IV. A method of observing root growth in the field; illustrated by observations in an irrigated apple orchard in British Columbia. Rep. East Malling Res. Stn. 1933, 86–91 (1934)

Rogers, W.S.: Root studies. VII.: A survey of the literature on root growth, with special reference to hardy fruit plants. J. Pomol. Hortic. Sci. *17*, 67–84 (1939a)

Rogers, W.S.: Root studies. VIII. Apple root growth in relation to rootstock, soil, seasonal and climatic factors. J. Pomol. Hortic. Sci. *17*, 99–130 (1939b)

Rogers, W.S.: Root studies. IX. The effect of light on growing apple roots: a trial with root observation boxes. J. Pomol. Hortic. Sci. *17*, 131–140 (1939c)

Rogers, W.S.: The East Malling root-observation laboratories. In: Root Growth. Whittington, W.J. (ed.). London: Butterworth, 1969, pp. 361–376

Rogers, W.S.: Die Wurzeln von Obstgewächsen. Obst Garten *90*, 296–299 (1971)

Rogers, W.S., Booth, G.A.: The roots of fruit trees. Sci. Hortic. *14*, 27–34 (1960)

Rogers, W.S., Head, G.C.: A new root-observation laboratory. Rep. East Malling Res. Stn. 1962, 55–57 (1963a)

Rogers, W.S., Head, G.C.: Studies of growing roots of fruit plants in a new underground root observation laboratory. Rep. 16th Int. Hortic. Congr. 1962, Brussels, Belgium. Vol. III, 311–318 (1963b)

Rogers, W.S., Head, G.C.: The roots of fruit plants. J. R. Hortic. Soc. *91*, 198–205 (1966)

Rogers, W.S., Head, G.C.: Studies of roots of fruit plants by observation panels and time-lapse photography. In: Methods of Productivity Studies in Root Systems and Rhizosphere Organisms. Int. Symp. USSR, 1968. Ed. by USSR Academy of Sciences, Leningrad: Nauka, 1968, pp. 176–185

Rogers, W.S., Head, G.C.: Factors affecting the distribution and growth of roots of perennial woody species. In: Root Growth. Whittington, W.J. (ed.). London: Butterworth, 1969, pp. 280–295

Rogers, W.S., Vyvyan, M.C.: The root systems of some ten year old apple trees on two different root stocks, and their relation to tree performance. Rep. East Malling Res. Stn. 1926-27, II. Suppl., 31–43 (1928)

Rogers, W.S., Vyvyan, M.C.: Root studies. V. Root stock and soil effect on apple root systems. J. Pomol. Hortic. Sci. *12*, 110–150 (1934)
Röhrig, E.: Die Wurzelentwicklung der Waldbäume in Abhängigkeit von den ökologischen Verhältnissen. Forstarchiv *37*, 217–229, 237–249 (1966)
Roo, De, H.C.: Root growth in Connecticut tobacco soils. Conn. Agric. Exp. Stn. Bull. No. 608 (1957) 36 pp
Roo, De, H.C.: Root development in coarse textured soils as related to tillage practices and soil compaction. Trans. 7th Int. Congr. Soil Sci., Madison, Wisc., USA, Vol. I, 622–628 (1960)
Roo, De, H.C.: Deep tillage and root growth. A study of tobacco growing in sandy loam soil. Conn. Agric. Exp. Stn. Bull. No. 644 (1961) 48 pp
Roo, De, H.C.: Root training by plastic tubes. II. Geotropic reaction of the roots of several species. Agron. J. *58*, 359–361 (1966a)
Roo, De, H.C.: Root training by plastic tubes. III. Soil aeration appraised by tube-grown plants. Agron. J. *58*, 483–486 (1966b)
Roo, De, H.C.: Tillage and root growth. In: Root Growth. Whittington, W.J. (ed.). London: Butterworth, 1969, pp. 339–357
Roo, De, H.C., Waggoner, P.E.: Root development of potatoes. Agron. J. *53*, 15–17 (1961)
Roo, De, H.C., Wiersum, L.K.: Root training by plastic tubes. Agron. J. *55*, 402–405 (1963)
Rotmistrov, V.G.: Root-system of cultivated plants of one year's growth. Odessa: South Russian Printing Co., 1909
Rotmistrov, V.G.: Das Wesen der Dürre. Dresden u. Leipzig: Th. Steinkopf, 1926
Rowse, H.R.: The effect of irrigation on the length, weight, and diameter of lettuce roots. Plant Soil *40*, 381–391 (1974)
Rowse, H.R., Phillips, D.A.: An instrument for estimating the total length of root in a sample. J. Appl. Ecol. *11*, 309–314 (1974)
Ruby, E.S., Young, V.A.: The influence of intensity and frequency of clipping on the root system of brownseed paspalum. J. Range Manage. *6*, 94–99 (1953)
Ruckenbauer, P.: Sortenspezifische Wurzel-Sproßverhältnisse bei Getreide in Nährlösungen. Bericht Arbeitstagung 1966 der Arbeitsgemeinschaft der Saatzuchtleiter in Gumpenstein, Österreich, 193–215 (1967a)
Ruckenbauer, P.: Untersuchungen über sortenspezifische Wurzel-Sproßverhältnisse bei Getreide in Nährlösungen. Diss. Hochsch. Bodenkultur Wien (1967b)
Ruckenbauer, P.: Die Beurteilung der Leistungsfähigkeit von Winterweizensorten nach bestimmten Organrelationen. Z. Acker- Pflanzenbau *130*, 273–290 (1969)
Rupprecht, H.: Rätselhaftes Wurzelwachstum? - Ein Beitrag zum "Blumentopfklima". Dtsch. Gartenbau *1*, 6–11 (1954)
Russell, R.S.: Wurzelsysteme und Pflanzenernährung – einige neue Gesichtspunkte. Endeavour *29*, 60–66 (1970)
Russell, R.S.: Root systems and nutrition. In: P.F. Wareing, J.P. Cooper: Potential Crop Production. London: Heinemann, 1971, pp. 100–116
Russell, R.S.: Plant Root Systems: Their Function and Interaction with the Soil. London: Mc Graw Hill, 1977
Russell, R.S., Adams, S.N.: The removal of plant roots from soil for the estimation of their phosphate content. Plant Soil *5*, 223–225 (1954)
Russell, R.S., Ellis, F.B.: Estimation of the distribution of plant roots in soil. Nature (London) *217*, 582–583 (1968)
Sachs, J.: Über das Wachstum der Haupt- und Nebenwurzeln. Arb. Bot. Inst. Würzburg *3*, 395–477, 584–634 (1873)
Safford, L.O.: Effect of fertilization on biomass and nutrient content of fine roots in a beech-birch-maple stand. Plant Soil *40*, 349–363 (1974)
Safford, L.O.: Seasonal variation in the growth and nutrient content of yellow-birch replacement roots. Plant Soil *44*, 439–444 (1976)
Saíz del Río, J.F., Fernández, C.E., Bellavita, O.: Distribution of absorbing capacity of coffee roots determined by radioactive tracers. Proc. Am. Hortic. Sci. *77*, 240–244 (1961)

Salim, M.H., Todd, G.W., Schlehuber, A.M.: Root development of wheat, oats and barley under conditions of soil moisture stress. Agron. J. *57*, 603–607 (1965)

Salonen, M.: Investigations of the root positions of field crops in the soils of Finland. (Fin., Engl. Sum.). Acta Agral. Fenn. *70*, *1*, 1–91 (1949)

Samoilova, E.M.: The study of the tree root systems on sandy soils. In: Methods of the Productivity Studies in Root Systems and Rhizosphere Organisms. Int. Symp. USSR, 1968. Ed. by USSR Academy of Sciences, Leningrad: Nauka, 1968, pp. 195–200

Sanders, J.L., Brown, D.A.: A new fiber optic technique for measuring root growth of soybeans under field conditions. Agron. J. *70*, 1073–1076 (1978)

Santantonio, D., Hermann, R.K., Overton, W.S.: Root biomass studies in forest ecosystems. Pedobiologia *17*, 1–31 (1977)

Sator, C.: Untersuchungen zur Unterscheidung lebender und toter Wurzeln. Landbauforsch. Völkenrode *22*, 87–92 (1972)

Sator, C., Bommer, D.: Methodological studies to distinguish functional from non-functional roots of grassland plants. In: Integrated Experimental Ecology. Ellenberg, H. (ed.). (Ecological Studies *2*) Berlin-Heidelberg-New York: Springer, 1971, pp. 72–74

Sauerbeck, D., Johnen, B.: Der Umsatz von Pflanzenwurzeln im Laufe der Vegetationsperiode und dessen Beitrag zur "Bodenatmung". Z. Pflanzenernähr. Bodenkd. *139*, 315–328 (1976)

Sauerbeck, D., Johnen, B., Six, R.: Atmung, Abbau und Ausscheidungen von Weizenwurzeln im Laufe ihrer Entwicklung, Landwirtschaftl. Forschung, Sonderheft *32*/I, 49–58 (1976)

Sayre, J.D., Morris, V.H.: The lithium method of measuring the extent of corn root systems. Plant Physiol. *15*, 761–764 (1940)

Scheffer, F., Kickuth, R.: Sprühtank für die Pflanzenaufzucht. Z. Pflanzenernähr. Düng. Bodenkd. *105*, 97–101 (1964)

Schmidt, E.W.: Die Wurzelbildmethode. Angew. Bot. *16*, 1–9 (1934)

Schmidt-Vogt, H.: Wachstum und Wurzelentwicklung von Schwarzerlen verschiedener Herkunft. Allg. Forst-Jagdztg. *142*, 149–156 (1971)

Schneider, G.: Vegetationsversuche mit 88 Hafersorten. Landwirtsch. Jahrb. *42*, 767–833 (1912)

Schoch, O.: Untersuchungen über die Stockraumbewurzelung verschiedener Baumarten im Gebiet der oberschwäbischen Jung- und Altmoräne. In: Standort, Wald und Waldwirtschaft in Oberschwaben. Herausgeg. von der Arbeitsgemeinschaft Oberschwäbische Fichtenreviere. Stuttgart, 1964, pp. 93–148

Schreiber, M.: Beiträge zur Kenntnis des Wurzelsystems der Lärche und der Fichte. Zentralbl. Ges. Forstwes. *52*, 78–103 (1926)

Schröder, D.: Unterscheidungsmerkmale der Wurzeln unserer Wiesen- und Weidenpflanzen. Landwirtsch. Jahrb. *64*, 41–64 (1926)

Schröder, D.: Unterscheidungsmerkmale der Wurzeln einiger Moor- und Grünlandpflanzen nebst einem Schlüssel zu ihrer Bestimmung und einem Anhang für die Bestimmung einiger Rhizome. Arb. Moor-Ver. Stn. Bremen: Schünemann, 1952

Schropp, W.: Der Vegetationsversuch. 1. Die Methodik der Wasserkultur höherer Pflanzen. Methodenbuch Bd. IX. Radebeul u. Berlin: Neumann, 1951

Schubart, A.: Erfahrungen und Beobachtungen über die Wurzelbildung und Wurzeltiefe mehrer landwirthschaftlicher Culturpflanzen, wie über die Keimkraft einiger Samenkörner. Chem. Ackersmann *1*, 193–201 (1855)

Schubart, A.: Ueber die Wurzelbildung der Cerealien, beobachtet bei Ausspülungen derselben in ihren verschiedenen Lebensperioden. Leipzig: R. Hoffmann, 1857

Schultz, A.M., Biswell, H.H.: A method for photographing roots. J. For. *53*, 138 (1955)

Schultz, J.E.: Root development of wheat at the flowering stage under different cultural practices. Agric. Rec. *1*, 12–17 (1974)

Schultz, R.P.: Root development of intensively cultivated slash pine. Proc. Soil Sci. Soc. Am. *36*, 158–162 (1972)

Schultz-Lupitz, A.: Zwischenfruchtbau auf leichtem Boden. Arb. Dtsch. Landwirtschaftsges. 7, 1–86 (1895)

Schulze, B.: Studien über die Bewurzelung unserer Kulturpflanzen. In: Festschrift zum fünfzigjährigen Jubiläum der Agrikulturchemischen Versuchs- und Kontrollstation der Landwirtschaftskammer für die Provinz Schlesien zu Breslau. Breslau, 1906, pp. 67–95

Schulze, B.: Wurzelatlas. Darstellung natürlicher Wurzelbilder der Halmfrüchte in verschiedenen Stadien der Entwicklung. Berlin: Parey, 2 Teile, 1911/1914

Schulze, E., Mues, H.: Ertragsleistung, Pflanzenbestand und Bewurzelung einer Grasnarbe bei verschiedener Düngungsweise. Z. Acker- Pflanzenbau *112*, 141–160 (1961)

Schulze, G.: Eine Monolithmethode zur Wurzelgewinnung bei Luzerne. Arch. Acker- Pflanzenbau *15*, 1055–1061 (1971)

Schumacher, R., Fankhauser, F., Schläpfer, E.: Die Wurzelentwicklung bei Apfelbäumen und ihre Beeinflussung durch den Hemmstoff 2,2-Dimethylhydrazid der Bernsteinsäure (Alar). Schweiz. Z. Obst- Weinbau *107*, 438–452 (1971)

Schuster, C.E., Stephenson, R.E.: Soil moisture, root distribution and aeration as factors in nut production in Western Oregon. Oreg. Agric. Exp. Stn. Bull. No. 372 (1940) 32 pp

Schuster, J.L.: Root development of native plants under three grazing intensities. Ecology *45*, 63–70 (1964)

Schuster, J.L., Wasser, C.H.: Nail-Board method for root sampling. J. Range Manage. *17*, 85–87 (1964)

Schuurman, J.J.: Influence of soil density on root development and growth of oats. Plant Soil *22*, 352–374 (1965)

Schuurman, J.J.: Effect of supplemental fertilization on growth of oats with restricted root development. Z. Acker-Pflanzenbau *133*, 315–320 (1971a)

Schuurman, J.J.: Effects of density of top and subsoil on root and top growth of oats. Z. Acker- Pflanzenbau *134*, 185–199 (1971b)

Schuurman, J.J., Boer, De, J.J.: The developmental pattern of roots and shoots of oats under favourable conditions. Neth. J. Agric. Sci. *18*, 168–181 (1970)

Schuurman, J.J., Boer, De, J.J.: The effect of soil compaction at various depths on root and shoot growth of oats. Neth. J. Agric. Sci. *22*, 133–142 (1974)

Schuurman, J.J., Goedewaagen, M.A.J.: A new method for the simultaneous preservation of profiles and root systems. Plant Soil *6*, 373–381 (1955)

Schuurman, J.J., Goedewaagen, M.A.J.: Methods for the Examination of Root Systems and Roots. Wageningen: Pudoc 2nd Ed., 1971

Schuurman, J.J., Knot, L.: The estimation of amounts of roots in samples bound for root investigations. (Dutch, Engl. Sum.). Versl. Landbouwkd. Onderz. *63,14*, 1957 (31 pp)

Schuurman, J.J., Knot, L.: The effect of nitrogen on the root and shoot development of *Lolium multiflorum* var. *westerwoldicum*. Neth. J. Agric. Sci. *22*, 82–88 (1974)

Schwaar, J.: Wurzeluntersuchungen aus Niedermooren. Ber. Dtsch. Bot. Ges. *84*, 745–757 (1971)

Schwaar, J.: Lebende Wurzeln in Hoch- und Niedermooren. Telma *2*, 73–82 (1972)

Schwaar, J.: Wurzeluntersuchungen auf Moorböden. Telma *3*, 119–136 (1973)

Schwarz, F.: Die Wurzelhaare der Pflanzen. Untersuch. Bot. Inst. Tübingen *1*, 135–188 (1883)

Schwarz, R.: Beiträge zur Kenntnis des Rohrglanzgrases (*Phalaris arundinacea*). Landwirtsch. Jahrb. *80*, 909–945 (1934)

Scott, H.D., Oliver, L.R.: Field competition between tall morningglory and soybean. II. Development and distribution of root systems. Weed Sci. *24*, 454–460 (1976)

Scully, N.J.: Root distribution and environment in a maple-oak forest. Bot. Gaz. *103*, 492–517 (1942)

Seelhorst, von, C.: Beobachtungen über die Zahl und den Tiefgang der Wurzeln verschiedener Pflanzen bei verschiedener Düngung des Bodens. J. Landwirtsch. *50*, 91–104 (1902)

Seiler, L.: Über das Wurzelwachstum und eine Methode zur quantitativen Untersuchung des Einflusses von Wirkstoffen. Ber. Schweiz. Bot. Ges. *61*, 622–663 (1951)

Sekera, F.: Über den zeitlichen Verlauf der Nährstoffaufnahme und Wurzelausbildung bei Gerste. Z. Pflanzenernähr. Düng. Bodenkd. Teil B, *7*, 527–530 (1928)

Seshadri, C.R., Rao, M.B., Muhamed, S.V.: Studies on root development in groundnut. Indian J. Agric. Sci. *28*, 211–215 (1958)

Shalyt, M.S.: Methods of studying the morphology and ecology of the subterranean part of individual plants and of entire plant communities. (Russ.). In: Polevaya Geobotanika. Lavrenko, E.M., Korchagin, A.A. (eds.). Moscow-Leningr: Academy of Sciences of the USSR Press, 1960, Vol. 2, pp. 369–447

Shalyt, M.S., Zhivotenko, L.F.: Overground and underground parts of certain grass and dwarf semishrub phytocoenoses of the Crimean Jaila (mountain pastures) and the technique of their estimation. In: Methods of Productivity Studies in Root Systems and Rhizosphere Organisms. Int. Symp. USSR, 1968. Ed. by USSR Academy of Sciences, Leningrad: Nauka, 1968, pp. 204–208

Shamoot, S., McDonald, L., Bartholomew, W.V.: Rhizo-deposition of organic debris in soil. Proc. Soil Sci. Soc. Am. *32*, 817–820 (1968)

Shanks, J.B., Laurie, A.: A progress report of some rose root studies. Proc. Am. Soc. Hortic. Sci. *53*, 473–488 (1949)

Shannon, E.L.: The production of root hairs by aquatic plants. Am. Midl. Nat. *50*, 474–479 (1953)

Shaver, G.R., Billings, W.D.: Root production and root turnover in a wet tundra ecosystem, Barrow, Alaska. Ecology *56*, 401–409 (1975)

Shepperd, J.H.: Root systems of field crops. N. D. Agric. Exp. Stn. Bull. No. 64, 525–536 (1905)

Shimura, K.: Root research phytotron. Jpn. Agric. Res. Q. *5*, No. 4, 54–57 (1970)

Shoop, M.C.: Foliage and root growth of weeping lovegrass. Unpublished Ph.D. Dissertation. Colorado State Univ., Fort Collins, Colorado, USA, 1978

Sideris, C.P.: Container for the study of the behavior of individual roots. Plant Physiol. *7*, 173–174 (1932)

Siebert, A.: Ergrünungsfähigkeit von Wurzeln. Beih. Bot. Zentralbl. Abt. 1 *37*, 185–216 (1920)

Simon, W., Eich, D.: Probleme und Methoden der Wurzeluntersuchungen unter Berücksichtigung leichterer Böden. Z. Acker- Pflanzenbau *100*, 179–198 (1956)

Simon, W., Eich, D., Zajons, A.: Vorläufiger Bericht über Beziehungen zwischen Wurzelmenge und Vorfruchtwert bei verschiedenen Klee- und Grasarten als Hauptfrucht auf leichten Böden. Z. Acker- Pflanzenbau *104*, 71–88 (1957)

Singer, F.P., Hutnick, R.J.: Excavating roots with water pressure. J. For. *63*, 37–38 (1965)

Singh, J.S., Coleman, D.C.: A technique for evaluating functional root biomass in grassland ecosystems. Canad. J. Bot. *51*, 1867–1870 (1973)

Singh, J.S., Coleman, D.C.: Evaluation of functional root biomass and translocation of photoassimilated ^{14}C in a shortgrass prairie ecosystem. In: The Belowground Ecosystem: A Synthesis of Plant Associated Processes. Marshall, J.K. (ed.). Range Science Department Science Series No. 26, Colorado State University, Fort Collins, USA, 1977, pp. 123–131

Sivakumar, M.V.K., Taylor, H.M., Shaw, R.H.: Top and root relations of field-grown soybeans. Agron. J. *69*, 470–473 (1977)

Skirde, W.: Bewurzelung der Rasendecke mit Beispielen für Abhängigkeit und Beeinflussung. Rasen-Turf-Gazon *2*, 112–115 (1971)

Slavík, B.: Methods of Studying Plant Water Relations. Ecological Studies *9*. Berlin-Heidelberg-New York: Springer, 1974

Slavíková, J.: Die Beschleunigung des Auswaschens der Wurzeln bei quantitativer Rhizologie. (Czech, Germ. Sum.). Rostl. Výroba *14*, 991–992 (1968)

Slavoňovský, F.: Dynamik der mechanischen Eigenschaften der Wurzeln von *Dianthus serotinus* im Laufe der Vegetationsperiode 1968/69. Teil I u. II. Scr. Fac. Sci. Nat. Ujep Brunensis, Biol. 2–3,*6*, 37–68 (1976)

Smet, De, L.A.H., Schuurman, J.J., Boekel, P.: The rootability of clay soils in the Dollard area for wheat. (Dutch, Engl. Sum.). Boor Spade *18*, 163–176 (1972)

Smucker, A.J.M., Erickson, A.E.: An aseptic mist chamber system: a method for measuring root processes of peas. Agron. J. *68*, 59–62 (1976)

Snell, R.S.: Simple apparatus for measuring resistance to root lodging in sweet corn. Agron. J. *58*, 362 (1966)

Snow, L.M.: The development of root hairs. Bot. Gaz. *40*, 12–48 (1905)

Soileau, J.M.: Activity of barley seedling roots as measured by strontium uptake. Agron. J. *65*, 625–628 (1973)
Soileau, J.M., Mays, D.A., Khasawneh, F.E., Kilmer, V.J.: The rhizotron-lysimeter research facility at TVA, Muscle Shoals, Alabama. Agron. J. *66*, 828–832 (1974)
Sokolow, A.W.: Die Verteilung der Nährstoffe im Boden und der Pflanzenertrag. Berlin: Deutscher Bauernverlag, 1956
Soong, N.K., Pushparajah, E., Singh, M.M., Talibudeen, O.: Determination of active root distribution of *Hevea brasiliensis* using radioactive phosphorus. Int. Symp. Soil Fertil. Eval.: New Delhi , India. Proc. Vol. 1, 309–315 (1971)
Spahr, K.: Untersuchungen über die Standfestigkeit von Sommergerste. Z. Acker-Pflanzenbau *110*, 299–331 (1960)
Speidel, B., Weiss, A.: Untersuchungen zur Wurzelaktivität unter einer Goldhaferwiese. Angew. Bot. *48*, 137–154 (1974)
Spencer, K.: Tomato root distribution under furrow irrigation. Aust. J. Agric. Res. *2*, 118–125 (1951)
Sperry, T.M.: Root systems in Illinois prairie. Ecology *16*, 178–202 (1935)
Spirhanzl, J.: Ein Versuch zur Feststellung der Tiefe und Art der Entwicklung des Wurzelsystems der Kulturpflanzen. (Czech, Germ. Sum.). Sb. Cesk. Akad. Zemĕd. *11*, 298–303 (1936)
Sprague, H.B.: Root development of perennial grasses and its relation to soil conditions. Soil Sci. *36*, 189–209 (1933)
Staebler, G.R., Rediske, J.H.: Progress in developing a radioactive tracer technique for mapping roots of Douglas-fir. Proc. Soc. Am. For. Meeting, Salt Lake City, Utah, USA, 164–166 (1958)
Stankov, N.Z.: Apparatus and techniques for studying plant roots. (Soviet) Plant Physiol. (Engl. translation from Fiziologiya Rastenii) 7, 609–612 (1960)
Stansell, J.R., Klepper, B., Browning, V.D., Taylor, H.M.: Effect of root pruning on water relations and growth of cotton. Agron. J. *66*, 591–592 (1974)
Steinberg, B.: Der Einfluß der Bodenbearbeitung auf die Wurzelentwicklung der Reben. Mitt. Rebe Wein, Obstbau Früchteverwert. Hess. Forschungsanst. Geisenheim *22*, 303–312 (1972)
Steinberg, B.: Methoden und Ergebnisse von Wurzeluntersuchungen bei Pfropfreben in Normalanlagen im Rheingau. Landwirtsch. Forsch. Sonderheft 28/I, 112–123 (1973a)
Steinberg, B.: Untersuchungen über die vertikale Verteilung der Wurzelspitzen auf verschiedenen Standorten. Wein-Wissenschaft *28*, 57–83 (1973b)
Steinberg, E., Eisenbarth, H.J.: Die Beeinflussung des Wurzelwachstums durch mechanische und chemische Bodenbearbeitung. Wein-Wissenschaft *26*, 400–404 (1971)
Steineck, O.: Sproßbildung und Wurzelwachstum verschiedener Kulturpflanzen bei konstantem Angebot steigender Nährstoffe N und K. Bodenkultur *15*, 268–284 (1964)
Stellwag-Carion, F.: Zur zahlenmäßigen Erfassung und schematischen Darstellung des Wurzel- und Schoßbildes. Züchter *9*, 184–188 (1937)
Stephan, S.: Untersuchungen im Wurzelbereich mit bodenkundlicher Dünnschlifftechnik: Möglichkeiten und Schwierigkeiten. Ber. Dtsch. Bot. Ges. *84*, 739–743 (1971)
Steubing, L.: Beiträge zur Ökologie der Wurzelsysteme von Pflanzen des flachen Sandstrandes. Z. Naturforsch. *4*b, 114–123 (1949)
Steubing, L.: Untersuchungen über die Konkurrenzwirkung von Gehölzwurzeln auf Ackerkulturen. Der Einfluß von Eichen als Standbäume in Hecken. Plant Soil *7*, 1–25 (1956)
Steubing, L.: Wurzeluntersuchungen an Feldschutzhecken. Z. Acker- Pflanzenbau *110*, 332–341 (1960)
Stevenson, D.S.: Effective soil volume and its importance to root and top growth of plants. Can. J. Soil Sci. *47*, 163–174 (1967)
Stigter, De, H.C.: A versatile irrigation-type waterculture for root-growth studies. Z. Pflanzenphysiol. *60*, 289–295 (1969)
Stoddard, L.A.: How long do roots of grasses live? Science *81*, 544 (1935)
Stoeckeler, J.H., Kluender, W.A.: The hydraulic method of excavating the root systems of plants. Ecology *19*, 355–369 (1938)

Stolzy, L.H., Barley, K.P.: Mechanical resistance encountered by roots entering compact soils. Soil Sci. *105*, 297–301 (1968)
Stone, J.F., Mulkey, J.R.: Pliable, root-permeable layers for separation of portions of experimental plant root systems. Science *133*, 329 (1969)
Stone, L.R., Teare, I.D., Nickell, C. D., Mayaki, W.C.: Soybean root development and soil water depletion. Agron. J. *68*, 677–680 (1976)
Stout, B.B.: Studies of the root systems of deciduous trees. Black Rock Forest, Cambridge, Mass., USA, Bull. No. 15 (1956) 45 pp
Straub, R.: Experimentelle Untersuchung über die Abhängigkeit des Wurzelwachstumsbeginnes bei Aspensämlingen von der Luft- und Bodentemperatur. Schweiz. Z. Forstwes. *117*, 60–67 (1966)
Strydom, E.: A root study of onions in an irrigation trial. S. Afr. J. Agric. Sci. *7*, 593–601 (1964)
Stucker, R., Frey, K.J.: The root-system distribution patterns for five oat varieties. Proc. Iowa Acad. Sci. *67*, 98–102 (1960)
Subbiah, B.V., Katyal, J.C., Narasimham, R.L., Dakshinamurti, C.: Preliminary investigations on root distribution of high yielding wheat varieties. Int. J. Appl. Radiat. Isot. *19*, 385–390 (1968)
Summerfield, R.J., Minchin, F.R.: Root growth in cowpea [*Vigna unguiculata* (L.) Walp.] seedlings. Trop. Agric. *53*, 199–209 (1976)
Sutton, R.F.: Form and Development of Conifer Root Systems. Commonwealth Agricultural Bureaux, Oxford, England. Techn. Commun. No. 7 (1969) 131 pp
Sweet, A.T.: Soil profile and root penetration as indicators of apple production in the lake shore district of Western New York. US-Dept. Agric. Circ. No. 303 (1933) 29 pp
Sytnik, K.M., Kniga, N.M., Musatenko, L.I.: Root Physiology. (Russ., Engl. Sum.). Kiew, 1972, 356 pp
Szembek, J.: Der Einfluß einiger agrotechnischer Maßnahmen auf die Entwicklung des Wurzelsystems der Bastardluzerne *(Medicago media)*. Z. Acker- Pflanzenbau *104*, 215–224 (1957)
Taerum, R., Gwynne, M.D.: Methods for studying grass roots under East African conditions and some preliminary results. East Afr. Agric. For. J. *35*, 55–65 (1969)
Taubenhaus, J.J., Ezekiel, W.N.: Acid injury of cotton roots. Bot. Gaz. *92*, 430–435 (1931)
Taubenhaus, J.J., Ezekiel, W.N., Rea, H.E.: Strangulation of cotton roots. Plant Physiol. *6*, 161–166 (1931)
Taylor, H.M.: The rhizotron at Auburn, Alabama – a plant root observation laboratory. Auburn Univ. Agric. Exp. Stn. Circ. No. 171 (1969) 9 pp
Taylor, H.M.: Observing plant root growth. In: Biological Effects in the Hydrological Cycle. Proc. Third Int. Sem. Hydrol. Prof. Purdue University, West Lafayette, Indiana, USA 1971. West Lafayette (1972) pp. 163–172
Taylor, H.M., Böhm, W.: Use of acrylic plastic as rhizotron windows. Agron. J. *68*, 693–694 (1976)
Taylor, H.M., Burnett, E.: Influence of soil strength on the root-growth habits of plants. Soil Sci. *98*, 174–180 (1964)
Taylor, H.M., Gardner, H.R.: Relative penetrating ability of different plant roots. Agron. J. *52*, 579–581 (1960a)
Taylor, H.M., Gardner, H.R.: Use of wax substrates in root penetration studies. Proc. Soil Sci. Soc. Am. *24*, 79–81 (1960b)
Taylor, H.M., Gardner, H.R.: Penetration of cotton seedling taproots as influenced by bulk density, moisture content, and strength of soil. Soil Sci. *96*, 153–156 (1963)
Taylor, H.M., Klepper, B.: Water uptake by cotton roots during an irrigation cycle. Aust. J. Biol. Sci. *24*, 853–859 (1971)
Taylor, H.M., Klepper, B.: Rooting density and water extraction patterns for corn (*Zea mays* L.). Agron. J. *65*, 965–968 (1973)
Taylor, H. M., Klepper, B.: Water relations of cotton. I. Root growth and water use as related to top growth and soil water content. Agron. J. *66*, 584–588 (1974)
Taylor, H.M., Klepper, B.: Water uptake by cotton root systems: An examination of assumptions in the single root model. Soil Sci. *120*, 57–67 (1975)

Taylor, H.M., Lund, Z.F.: The root system of corn. Proc. 25th Annu. Corn Sorghum Res. Conf. *25*, 175–179 (1970)

Taylor, H.M., Burnett, E., Welch, N.H.: Cotton growth and yield as affected by taproot diameter within a simulated restraining soil layer. Agron. J. *55*, 143–144 (1963)

Taylor, H.M., Burnett, E., Booth, G.D.: Taproot elongation rates of soybeans. Z. Acker-Pflanzenbau *146*, 33–39 (1978)

Taylor, H.M., Huck, M.G., Klepper, B., Lund, Z.F.: Measurement of soil-grown roots in a rhizotron. Agron. J. *62*, 807–809 (1970)

Ten Eyck, A.M.: A study of the root systems of wheat, oats, flax, corn, potatoes and sugar beets, and of the soil in which they grew. N. D. Agric. Exp. Stn. Bull. No. *36*, 333–362 (1899)

Ten Eyck, A.M.: A study of the root systems of cultivated plants grown as farm crops. N. D. Agric. Exp. Stn. Bull. No. *43*, 535–560 (1900)

Ten Eyck, A.M.: The roots of plants. Kansas Agric. Exp. Stn. Bull. No. *127*, 199–252 (1904)

Tennant, D.: A test of a modified line intersect method of estimating root length. J. Ecol. *63*, 995–1001 (1975)

Tennant, D.: Root growth of wheat. I. Early patterns of multiplication and extension of wheat roots including effects of levels of nitrogen, phosphorus and potassium. Aust. J. Agric. Res. *27*, 183–196 (1976)

Thangavelu, S., Meenakshi, K., Rajagopal, C.K.: Studies on drought resistance in sorghum-root characters. Madras Agric. J. *56*, 64–67 (1969)

Tharp, B.C., Muller, C.H.: A rapid method for excavating root systems of native plants. Ecology *21*, 347–350 (1940)

Thiel, H.: Ueber die Bewurzelung einiger unserer Culturpflanzen. Z. Landwirthsch. Ver. Großherzogthums Hessen. 323–327, 332–336, (No. 37) (1870)

Thiel, H.: Anleitung zu Wurzelstudien. Mitt. Dtsch. Landwirtschaftsges. *7*, 75–76 (1892)

Thomas, A.S.: Observations on the root-systems of *Robusta* coffee and other tropical crops in Uganda. Emp. J. Exp. Agric. *12*, 191–206 (1944)

Thomas, R.: A method of studying the roots of sugarcane. Agric. J. India *22*, 138–142 (1927)

Thompson, D.L.: Field evaluation of corn root clumps. Agron. J. *60*, 170–172 (1968)

Thorup, R.M.: Root development and phosphorus uptake by tomato plants under controlled soil moisture conditions. Agron. J. *61*, 808–811 (1969)

Tornau, O., Stölting, H.: Untersuchungen über die Bewurzelung der Ackerbohne. J. Landwirtsch. *90*, 1–32 (1944)

Torrey, J.G., Clarkson, D.T. (eds.): The Development and Function of Roots. Third Cabot Symposium. London-New York-San Francisco: Academic Press, 1975, 618 pp

Torssell, B.W.R., Begg, J.E., Rose, C. W., Byrne, F.G.: Stand morphology of Townsville lucerne (*Stylosanthes humilis*). Seasonal growth and root development. Aust. J. Exp. Agric. Anim. Husb. *8*, 533–543 (1968)

Troughton, A.: The application of the allometric formula to the study of the relationship between the roots and shoots of young grass plants. Agric. Progr. *30*, 59–65 (1955)

Troughton, A.: Studies on the growth of young grass plants with special reference to the relationship between the root and shoot systems. J. Br. Grassl. Soc. *11*, 56–65 (1956)

Troughton, A.: The Underground Organs of Herbage Grasses. Commonw. Bur. Pastures Field Crops, Hurley, Berkshire, England. Bull. No. 44 (1957) 163 pp

Troughton, A.: Further studies on the relationship between the shoot and root systems of grasses. J. Br. Grassl. Soc. *15*, 41–47 (1960)

Troughton, A.: The Roots of Temperate Cereals. Commonw. Bur. Pastures Field Crops, Hurley, Berkshire, England. Mimeographed Publication No. 2 (1962) 91 pp

Troughton, A.: A comparison of five varieties of *Lolium perenne* with special reference to the relationship between the root and shoot systems. Euphytica *12*, 49–56 (1963)

Troughton, A.: Grass roots, their morphology and relationship with the shoot system. In: Methods of the Productivity Studies in Root Systems and Rhizosphere Organisms. Symp. USSR, 1968. Ed. by USSR Academy of Sciences, Leningrad: Nauka, 1968, pp. 213–220

Troughton, A.: The growth and function of the root in relation to the shoot. In: Structure and Function of Primary Root Tissues. Bratislava: Veda, 1974, pp. 153–164

Troughton, A.: Relationship between the root and shoot systems of grasses. In: The Belowground Ecosystem: A Synthesis of Plant-Associated Processes. Marshall, J.K. (ed.). Range Science Department Science Series No. 26, Colorado State University, Fort Collins, USA, 1977, pp. 39–51

Troughton, A., Whittington, W.J.: The significance of genetic variation in root systems. In: Root Growth. Whittington, W.J. (ed.). London: Butterworth, 1969, pp. 296–314

Trouse, A.C., Humbert, R.P.: Some effects of soil compaction on the development of sugar cane roots. Soil Sci. *91*, 208–217 (1961)

Trouse, A.C., Parish, D.H., Taylor, H.M.: Soil conditions as they affect plant establishment, root development, and yield. In: Compaction of Agricultural Soils. ASAE Monograph published by the Am. Soc. Agric. Eng. St. Joseph, Michigan, USA, 1971, pp. 225–312

Tukey, H.B., Brase, K.D.: Studies of top and root growth of young apple trees in soil and peat-soil mixtures of varying moisture contents. Proc. Am. Soc. Hortic. Sci. *36*, 18–27 (1938)

Turner, L.M.: A comparison of roots of Southern shortleaf pine in three soils. Ecology *17*, 649–658 (1936)

Ueno, M., Yoshihara, K., Okada, T.: Living root system distinguished by the use of carbon-14. Nature (London) *213*, 530–532 (1967)

Upchurch, R.P.: The use of trench-wash and soil-elution methods for studying alfalfa roots. Agron. J. *43*, 552–555 (1951)

USSR Academy of Sciences: Methods of Productivity Studies in Root Systems and Rhizosphere Organisms. Int. Symp. USSR, 1968. Leningrad: Nauka, 1968, 240 pp

Vater, H.: Die Bewurzelung der Kiefer, Fichte und Buche. Tharandter Forstl. Jahrb. *78*, 65–85 (1927)

Vávra, M.: Simplified method for studies of root growth dynamics in fruit trees. (Czech, Engl. Sum.). Acta Univ. Agric. (Brno), Fac. Agron. *23*, 85–97 (1975)

Veihmeyer, F.J.: An improved soil-sampling tube. Soil Sci. *27*, 147–152 (1929)

Veihmeyer, F.J., Hendrickson, A.H.: Soil moisture as an indication of root distribution in deciduous orchards. Plant Physiol. *13*, 169–177 (1938)

Veihmeyer, F.J., Hendrickson, A.H.: Soil density and root penetration. Soil Sci. *65*, 487–493 (1948)

Venkatraman, R.S., Thomas, R.: Simple contrivances for studying root development in agricultural crops. Agric. J. India *19*, 509–514 (1924)

Vetter, H., Früchtenicht, K.: Die Bewurzelung einiger Pflanzenarten in verschiedenen Entwicklungsstadien und bei gestaffelter Phosphatdüngung. Phosphorsäure *28*, 1–18 (1969)

Vetter, H., Scharafat, S.: Die Wurzelverbreitung landwirtschaftlicher Kulturpflanzen im Unterboden. Z. Acker- Pflanzenbau *120*, 275–298 (1964)

Vijayalakshmi, K., Dakshinamurti, C.: Quantitative estimation of root weight using ^{32}P plant injection technique. J. Nucl. Agric. Biol. *4*, 98–100 (1975)

Vijayalakshmi, K., Dakshinamurti, C.: Limitations of the ^{32}P isotope injection technique for the study of the root systems of wheat, mung and cowpeas. Plant Soil *46*, 113–125 (1977)

Voorhees, W.B.: Root elongation along a soil-plastic container interface. Agron. J. *68*, 143 (1976)

Waddington, J.: Observation of plant roots in situ. Can. J. Bot. *49*, 1850–1852 (1971)

Wadleigh, C.H., Gauch, H.G., Strong, D.G.: Root penetration and moisture extraction in saline soil by crop plants. Soil Sci. *63*, 341–349 (1947)

Wagenhoff, A.: Untersuchungen über die Entwicklung des Wurzelsystems der Kiefer auf diluvialen Sandböden. Z. Forst- Jagdwes. *70*, 449–494 (1938)

Wagenknecht, E.: Wurzeluntersuchungen und ihre Bedeutung für standortgerechten Waldbau. Arch. Forstwes. *4*, 397–406 (1955)

Wagner, P.: Ueber Land- und Wasserwurzeln. J. Landwirtsch. *18*, 103–110 (1870)

Walker, M.E., Stansell, J.R. Jr., Shannon, J.E.: Precision soil sampler. Agron. J. *68*, 431–432 (1976)
Wallace, A., Bamberg, S.A., Cha, J.W., Romney, E.M.: Partitioning of photosynthetically fixed ^{14}C in perennial plants in the Northern Mojave Desert. In: The Belowground Ecosystem: A Synthesis of Plant-Associated Processes. Marshall, J.K. (ed.). Range Science Department Science Series No. 26, Colorado State University, Fort Collins, USA, 1977, pp. 141–148
Ward, K.J., Klepper, B., Rickman, R.W., Allmaras, R.R.: Quantitative estimation of living wheat-root lengths in soil cores. Agron. J. *70*, 675–677 (1978)
Warembourg, F.R., Paul, E.A.: Seasonal transfers of assimilated ^{14}C in grassland: plant production and turnover, translocation and respiration. In: The Belowground Ecosystem: A Synthesis of Plant-Associated Processes. Marshall, J.K. (ed.). Range Science Department Science Series No. 26, Colorado State University, Fort Collins, USA, 1977, pp. 133–140
Warsi, A.S., Wright, B.C.: Influence of nitrogen on the root growth of grain sorghum. Indian J. Agric. Sci. *43*, 142–147 (1973)
Weaver, J.E.: A study of the root-systems of prairie plants of Southeastern Washington. Plant World *18*, 227–248, 273–292 (1915)
Weaver, J.E.: The Ecological Relations of Roots. Carnegie Institution of Washington. Publ. No. *286*, Washington D.C., 1919
Weaver, J.E.: Root Development in the Grassland Formation. Carnegie Institution of Washington. Publ. No. *292*, Washington D.C., 1920
Weaver, J.E.: Investigations on the root habits of plants. Am. J. Bot. *12*, 502–509 (1925)
Weaver, J.E.: Root Development of Field Crops. New York-London: McGraw-Hill Book Co., 1926, 291 pp
Weaver, J.E.: Underground plant development in its relation to grazing. Ecology *11*, 543–557 (1930)
Weaver, J.E.: Classification of root systems of forbs of grassland and a consideration of their significance. Ecology *39*, 393–401 (1958a)
Weaver, J.E.: Summary and interpretation of underground development in natural grassland communities. Ecol. Monograph. *28*, 55–78 (1958b)
Weaver, J.E.: The living network in prairie soils. Bot. Gaz. *123*, 16–28 (1961)
Weaver, J.E., Albertson, F.W.: Resurvey of grasses, forbs, and underground plant parts at the end of the great drought. Ecol. Monograph. *13*, 63–117 (1943)
Weaver, J.E., Bruner, W.E.: Root Development of Vegetable Crops. New York-London: McGraw-Hill Book Co., 1927, 351 pp
Weaver, J.E., Christ, J.W.: Relation of hardpan to root penetration in the Great Plains. Ecology *3*, 237–249 (1922)
Weaver, J.E., Clements, F.E.: Plant Ecology. New York-London: McGraw-Hill Book Co., 2nd ed., 1938, 601 pp
Weaver, J.E., Darland, R.W.: A method of measuring vigor of range grasses. Ecology *28*, 147–162 (1947)
Weaver, J.E., Darland, R.W.: Quantitative study of root systems in different soil types. Science *110*, 164–165 (1949a)
Weaver, J.E., Darland, R.W.: Soil-root relationship of certain native grasses in various soil types. Ecol. Monograph. *19*, 303–338 (1949b)
Weaver, J.E., Himmel, W.J.: Relation between the development of root system and shoot under long- and short-day illumination. Plant Physiol. *4*, 435–457 (1929)
Weaver, J.E., Himmel, W.J.: Relation of increased water content and decreased aeration to root development in hydrophytes. Plant Physiol. *5*, 69–92 (1930)
Weaver, J.E., Kramer, J.: Root system of *Quercus macrocarpa* in relation to the invasion of prairie. Bot. Gaz. *94*, 51–85 (1932)
Weaver, J.E., Voigt, J.W.: Monolith method of root-sampling in studies on succession and degeneration. Bot. Gaz. *111*, 286–299 (1950)
Weaver, J. E., Zink, E.: Extent and longevity of the seminal roots of certain grasses. Plant Physiol. *20*, 359–379 (1945)

Weaver, J. E., Jean, F.C., Christ, J.W.: Development and Activities of Roots of Crop Plants. A Study in Crop Ecology. Carnegie Institution of Washington. Publ. No. *316*, Washington D.C., 1922

Weaver, J.E., Kramer, J., Reed, M.: Development of root and shoot of winter wheat under field environment. Ecology *5*, 26–50 (1924)

Webb, W.L.: Rates of current photosynthate accumulation in roots of Douglas fir seedlings: seasonal variation. In: The Belowground Ecosystem: A Synthesis of Plant-Associated Processes. Marshall, J. K. (ed.). Range Science Department Science Series No. 26, Colorado State University, Fort Collins, USA, 1977, pp. 149–152

Weihing, R.M.: The comparative root development of regional types of corn. J. Am. Soc. Agron. *27*, 526–537 (1935)

Weir, L.C.: The use of compressed air to excavate roots of forest trees. Can. Dep. For. Rural Dev. BiMon. Res. Notes. Vol. 22, No. 6, p. 2 (1966)

Weiske, H.: Über die Zusammensetzung und Menge der dem Acker nach der Ernte verbleibenden Stoppel- und Wurzelrückstände. Landwirtsch. Vers. Stn. *14*, 105–119 (1871)

Welbank, P.J., Williams, E.D.: Root growth of barley crop estimated by sampling with portable powered soil-coring equipment. J. Appl. Ecol. *5*, 477–481 (1968)

Welbank, P.J., Gibb, M.J., Taylor, P.J., Williams, E.D.: Root growth of cereal crops. In: Report Rothamsted Exp. Stn. 1973, Part 2, pp. 26–66 (1974)

Weller, F.: Vergleichende Untersuchungen über die Wurzelverteilung von Obstbäumen in verschiedenen Böden des Neckarlandes. Arb. Landwirtsch. Hochsch. Hohenheim *31*, (1964) 181 pp

Weller, F.: Die Ausbreitung der Pflanzenwurzeln im Boden in Abhängigkeit von genetischen und ökologischen Faktoren. Eine Literaturauswertung unter besonderer Berücksichtigung der Obstgehölze. Arb. Landwirtsch. Hochsch. Hohenheim *32* (1965) 123 pp

Weller, F.: Zur Durchwurzelbarkeit von Tonschiefern und Tonmergeln. Erwerbsobstbau *8*, 121–124 (1966a)

Weller, F.: Horizontale Saugwurzelverteilung und Düngerausbringung in Apfelanlagen. Erwerbsobstbau *8*, 181–184 (1966b)

Weller, F.: Zur zeitlichen Variabilität der Dichte des Saugwurzelbesatzes von Apfelbäumen. Erwerbsobstbau *9*, 168–170 (1967)

Weller, F.: Zur Beeinflussung der vertikalen Saugwurzelverteilung von Apfelbäumen durch Bodenpflegemaßnahmen. Erwerbsobstbau *10*, 4–6 (1968)

Weller, F.: A method for studying the distribution of absorbing roots of fruit trees. Exp. Agric. *7*, 351–361 (1971)

Werenfels, L.: Wurzelentwicklung einer Apfelunterlage in Abhängigkeit von der Bodenfeuchtigkeit. Berichte Schweiz. Bot. Ges. *77*, 5–48 (1967)

Werner, D., Schultz-Lupitz, A.: Über die Bewurzelung der landwirtschaftlichen Kulturgewächse und deren Bedeutung für den praktischen Ackerbau. Jahrb. Dtsch. Landwirtschaftsges. *6*, Teil II, 71–90 (1891)

Werner, O.: Die Maispflanze auf einem trockenharten Wurzelfaden voll wachsend. Biol. Gen. *3*, 689–709 (1931)

Wetzel, M.: Der Futterroggen, die Futterroggengemische und sonstige Futtergemische des Winterzwischenfruchtbaues als Lieferanten von Futter und Wurzeln. Wiss. Z. Univ. Rostock, Math.-Nat. Reihe *7*, 89–166, 307–366 (1957/58)

Wetzel, M.: Die Durchporung und Durchwurzelung von Dauerweidestandorten der Altmärkischen Wische. Z. Landeskultur *1*, 147–164 (1960a)

Wetzel, M.: Untersuchungen über Bewurzelungsverhältnisse von *Agrostis alba* f. *salina*. Ein Beitrag zur Trittfestigkeit der Weidenarbe. Z. Landeskultur *1*, 190–201 (1960b)

Whitcomb, C.E.: Influence of tree root competition on growth response of four cool season turfgrasses. Agron. J. *64*, 355–359 (1972)

Whitcomb, C.E., Roberts, E.C., Landers, R.Q.: A connecting pot technique for root competition investigations between woody plants or between woody and herbaceous plants. Ecology *50*, 326–328 (1969)

Whittington, W.J. (ed.): Root Growth. Proc. 15th Easter School in Agric. Sci. University Nottingham, 1968. London: Butterworth, 1969, 450 pp
Wiersma, D.: The soil environment and root development. Adv. Agron. *11*, 43–51 (1959)
Wiersum, L.K.: The relationship of the size and structural rigidity of pores to their penetration by roots. Plant Soil *9*, 75–85 (1957)
Wiersum, L.K.: Density of root branching as affected by substrate and separate ions. Acta Bot. Neerl. *7*, 174–190 (1958)
Wiersum, L.K.: Potential subsoil utilization by roots. Plant Soil *27*, 383–400 (1967a)
Wiersum, L.K.: Root-system development. In: Soil-Moisture and Irrigation Studies. Panel Proceedings Series Ed. by the Int. A. E. A., Vienna, 1967b, pp. 83–96
Wiese, A.F.: Rate of weed root elongation. Weed Sci. *16*, 11–13 (1968)
Wilde, S.A., Voigt, G.K.: Absorption-transpiration quotient of nursery stock. J. For. *47*, 643–645 (1949)
Willard, C.J., McClure, G.M.: The quantitative development of tops and roots in bluegrass with an improved method of obtaining root yields. J. Am. Soc. Agron. *24*, 509–514 (1932)
Willatt, S.T., Struss, R.G., Taylor, H.M.: In situ root studies using neutron radiography. Agron. J. *70*, 581–586 (1978)
Williams, T. E., Baker, H.K.: Studies on the root development of herbage plants. I. Techniques of herbage root investigations. J. Br. Grassl. Soc. *12*, 49–55 (1957)
Wilson, B.F.: Structure and growth of woody roots of *Acer rubrum* L. Harvard For. Pap. (Harvard Univ. Petersham, Mass., USA) No. 11, 1–14 (1964)
Wilson, B.F.: Distribution of secondary thickening in tree root systems. In: The Development and Function of Roots. Ed. by J.G. Torrey, D.T. Clarkson. London-New York-San Francisco: Academic Press, 1975, pp. 197–219
Wilson, B.F., Horsley, S.B.: Ontogenetic analysis of tree roots in *Acer rubrum* and *Betula papyrifera*. Am. J. Bot. 57, 161–164 (1970)
Wind, G.P.: Root growth in acid soils. Neth. J. Agric. Sci. *15*, 259–266 (1967)
Witte, K.: Beitrag zu den Grundlagen des Grasbaus. Landwirtsch. Jahrb. *69*, 253–310 (1929)
Woods, F.W.: Slash pine roots start growth soon after planting. J. For. *57*, 209 (1959)
Woods, F.W., Brock, K.: Interspecific transfer of Ca-45 and P-32 by root systems. Ecology *45*, 886–889 (1964)
Worzella, W.W.: Root development in hardy and non-hardy winter wheat varieties. J. Am. Soc. Agron. *24*, 626–637 (1932)
Yaalon, D.H.: "Calgon" no longer suitable. J. Soil Sci. Soc. Am. *40*, 333 (1976)
Yeatman, C.W.: Tree Root Development on Upland Heaths. For. Comm. Bull. No. 21. London: Her Majesty's Stationary Office (1955) 72 pp
Yli-Vakkuri, P.: Untersuchungen über organische Wurzelverbindungen zwischen Bäumen in Kiefernbeständen. (Fin., Germ. Sum.). Acta For. Fenn. *60,3*, 1–117 (1953)
Yocum, W.W.: Root development of young delicious apple trees as affected by soils and by cultural treatments. University of Nebraska. Agric. Exp. Stn. Res. Bull. No. 95 (1937) 55 pp
Yorke, J.S.: A review of techniques for studying root systems. For. Res. Lab. St. John's, New Foundland, Canada. Inf. Rep. N-X-20, (1968) 29 pp
Younis, A.F., Hatata, M.A.: Techniques to obtain uniformly growing young roots for the study of salt effects. Plant Soil *34*, 49–56 (1971)
Zapryagaeva, V.I.: Root systems of dominant and edificator tree species in the Pamir-Aley forest vegetation. In: Methods of Productivity Studies in Root Systems and Rhizosphere Organisms. Int. Symp. USSR, 1968. Ed. by USSR Academy of Sciences, Leningrad: Nauka, 1968, pp. 234–238
Zillmann, K.-H.: Beobachtung des Wurzelwachstums in Feldbeständen. Dtsch. Landwirtsch. *7*, 394–400 (1956)
Zimmermann, R.P., Kardos, L.T.: Effect of bulk density on root growth. Soil Sci. *91*, 280–289 (1961)

Zobel, R.W.: The genetics of root development. In: The Development and Function of Roots. Ed. by J.G. Torrey, D.T. Clarkson. London-New York-San Francisco: Academic Press, 1975, pp. 261–275

Zopf, P.E., Nettles, V.F.: Root development of contender snap beans. Proc. Fl. State Hortic. Soc. *68*, 175–177 (1955)

Zöttl, H.: Düngung und Feinwurzelverteilung in Fichtenbeständen. Mitt. Staatsforstverwaltung Bayern *34*, 333–342 (1964)

Zuber, M.S.: Evaluations of corn root systems under various environments. Rep. 23rd Corn Sorghum Res. Conf. Chicago, 67–75 (1968)

Subject Index

M. Kluge, I. P. Ting

Crassulacean Acid Metabolism

Analysis of an Ecological Adaptation

1978. 112 figures, 21 tables. XII, 210 pages
(Ecological Studies, Volume 30)
ISBN 3-540-08979-9

The acid metabolism of Crassulaceae and other succulents, known today as Crassulacean Acid Metabilism (CAM), has developed as an adaptation of plants to arid zones. CAM, which has fascinated plant physiologists for more than 150 years, is understood to be a modification of photosynthetic carbon assimilation, similar to C_4 synthesis. This book represents the first comprehensive treatment of this phenomena and covers taxonomic, morphological-anatomic, biochemical, physiological, and ecologcal aspects. The work discusses also the significance of CAM plants in agriculture and technology.

W. Tranquillini

Physiological Ecology of the Alpine Timberline

Tree Existence in High Altitudes with Special Reference to the European Alps
Translated from the German by U. Benecke

1979. 67 figures, 21 tables. Approx. 150 pages
(Ecological Studies, Volume 31)
ISBN 3-540-09065-7

In all the Earth's mountain areas, nature sets an upper limit to forest growth – the timberline. With the advent of ecophysiological field methods in 1930, the study of this extremely complex phenomenon made rapid strides. Never has there been a more comprehensive analysis of the struggle of trees for survival at their highest altitute of subsistence. Based on decades of research work at various timberline stands in the Austrian Alps, this book explores in-depth how climate changes with increasing altitute, how trees adapt to this increase, and at what point adaption becomes impossible. The book offers the botanist new insights into the formation of one of the most important vegetation boundaries and provides forestry and agriculture with a sound scientific basis for the preservation and restoration of protective forest in mountain areas.

Perspectives in Grassland Ecology

Results and Applications of the US/IBP Grassland Biome Study
Editor: N. R. French

1979. 58 figures. XVII, 686 pages
(Ecological Studies, Volume 32)
ISBN 3-540-90384-4

Perspectives in Grassland Ecology examines the biomass structure of grasslands in North America and other areas of the world. It compares trophic structure, and discribes structure modifications occurring when water and nutrients are artificially supplied. The most important driving variables, processes, and the major functional aspects of production are analyzed and modeled mathematically. Unique in its detailed comparative analysis of the major grassland types, this book will be appreciated by researchers in range science, ecology, agronomy and botany.

W. Larcher

Physiological Plant Ecology

Translator from the German: M. A. Biederman-Thorson

1975. 152 figures, 40 tables. XIV, 252 pages
ISBN 3-540-07336-1

"... The present book, written by a wellknown and respected pioneer in this field, is a translation of the German edition 'Ökologie der Pflanzen' first published in 1973, but with certain additions and corrections.
The arrangement of the eight chapters follows along more or less conventional lines. ... The book contains an enormous amount of factual material, including accounts of the underlying mechanisms of photosynthesis, ion and water uptake and transport; it is profusely illustrated with clear and informative diagrams and tables... the translation is excellent and the book as a whole reads quite well. ..."

The New Phytologist

Springer-Verlag Berlin Heidelberg New York